Mining Structures of Factual Knowledge from Text

An Effort-Light Approach

Synthesis Lectures on Data Mining and Knowledge Discovery

Editors
Jiawei Han, *University of Illinois at Urbana-Champaign*
Lise Getoor, *University of California, Santa Cruz*
Wei Wang, *University of California, Los Angeles*
Johannes Gehrke, *Cornell University*
Robert Grossman, *University of Chicago*

Synthesis Lectures on Data Mining and Knowledge Discovery is edited by Jiawei Han, Lise Getoor, Wei Wang, Johannes Gehrke, and Robert Grossman. The series publishes 50- to 150-page publications on topics pertaining to data mining, web mining, text mining, and knowledge discovery, including tutorials and case studies. Potential topics include: data mining algorithms, innovative data mining applications, data mining systems, mining text, web and semi-structured data, high performance and parallel/distributed data mining, data mining standards, data mining and knowledge discovery framework and process, data mining foundations, mining data streams and sensor data, mining multi-media data, mining social networks and graph data, mining spatial and temporal data, pre-processing and post-processing in data mining, robust and scalable statistical methods, security, privacy, and adversarial data mining, visual data mining, visual analytics, and data visualization.

Mining Structures of Factual Knowledge from Text: An Effort-Light Approach
Xiang Ren and Jiawei Han
2018

Individual and Collective Graph Mining: Principles, Algorithms, and Applications
Danai Koutra and Christos Faloutsos
2017

Phrase Mining from Massive Text and Its Applications
Jialu Liu, Jingbo Shang, and Jiawei Han
2017

Exploratory Causal Analysis with Time Series Data
James M. McCracken
2016

Mining Structures of Factual Knowledge from Text: An Effort-Light Approach

Xiang Ren and Jiawei Han

ISBN: 978-3-031-00784-2   paperback
ISBN: 978-3-031-01912-8   ebook
ISBN: 978-3-031-00107-9   hardcover

DOI 10.1007/978-3-031-01912-8

A Publication in the Springer series
SYNTHESIS LECTURES ON DATA MINING AND KNOWLEDGE DISCOVERY

Lecture #15
Series Editors: Jiawei Han, *University of Illinois at Urbana-Champaign*
               Lise Getoor, *University of California, Santa Cruz*
               Wei Wang, *University of California, Los Angeles*
               Johannes Gehrke, *Cornell University*
               Robert Grossman, *University of Chicago*
Series ISSN
Print 2151-0067   Electronic 2151-0075

Mining Structures of
Factual Knowledge from Text

An Effort-Light Approach

Xiang Ren
University of Southern California

Jiawei Han
University of Illinois at Urbana-Champaign

SYNTHESIS LECTURES ON DATA MINING AND KNOWLEDGE DISCOVERY #15

ABSTRACT

The real-world data, though massive, is largely unstructured, in the form of natural-language text. It is challenging but highly desirable to mine structures from massive text data, without extensive human annotation and labeling. In this book, we investigate the principles and methodologies of mining structures of factual knowledge (e.g., entities and their relationships) from massive, unstructured text corpora.

Departing from many existing structure extraction methods that have heavy reliance on human annotated data for model training, our effort-light approach leverages human-curated facts stored in external knowledge bases as distant supervision and exploits rich data redundancy in large text corpora for context understanding. This effort-light mining approach leads to a series of new principles and powerful methodologies for structuring text corpora, including: (1) entity recognition, typing, and synonym discovery; (2) entity relation extraction; and (3) open-domain attribute-value mining and information extraction. This book introduces this new research frontier and points out some promising research directions.

KEYWORDS

mining factual structures, information extraction, knowledge bases, entity recognition and typing, relation extraction, entity synonym mining, distant supervision, effort-light approach, classification, clustering, real-world applications, scalable algorithms

To my wonderful parents for their love and support.

– Xiang Ren

To my wife Dora, son Lawrence, and grandson Emmett for their love.

– Jiawei Han

Contents

6   Synonym Discovery from Large Corpus 75

Meng Qu

Department of Computer Science, University of Illinois at Urbana-Champaign

PART II   Extracting Typed Relationships 85

7   Joint Extraction of Typed Entities and Relationships 87

8   Pattern-Enhanced Embedding Learning for Relation Extraction 111

Meng Qu

Department of Computer Science, University of Illinois at Urbana-Champaign

9 Heterogeneous Supervision for Relation Extraction 119

Liyuan Liu

Department of Computer Science, University of Illinois at Urbana-Champaign

10 Indirect Supervision: Leveraging Knowledge from Auxiliary Tasks 129

Zeqiu Wu

Department of Computer Science, University of Illinois at Urbana-Champaign

Acknowledgments

The authors would like to acknowledge Wenqi He, Liyuan Liu, Meng Qu, Ellen Wu, Qi Zhu, Jingbo Shang, and Meng Jiang for their tremendous research collaborations.

Han's work was supported in part by U.S. Army Research Lab. under Cooperative Agreement No. W911NF-09-2-0053 (NSCTA), DARPA under Agreement No. W911NF-17-C-0099, National Science Foundation IIS 16-18481, IIS 17-04532, and IIS-17-41317, DTRA HDTRA11810026, and grant 1U54GM114838 awarded by NIGMS through funds provided by the trans-NIH Big Data to Knowledge (BD2K) initiative (www.bd2k.nih.gov).

Ren's work was sponsored by Google PhD Fellowship, ACM SIGKDD Scholarship, and Richard T. Cheng Endowed Fellowship.

Any opinions, findings, and conclusions or recommendations expressed in this document are those of the author(s) and should not be interpreted as the views of any funding agencies.

Xiang Ren and Jiawei Han
June 2018

CHAPTER 1

Introduction

The success of data mining technology is largely attributed to the efficient and effective analysis of structured data. However, the majority of existing data generated in our computerized society is unstructured or loosely structured, and is typically "text-heavy." People are soaked with vast amounts of natural-language text data, ranging from news articles, social media posts, online advertisements, and scientific publications, to a wide range of textual information from various domains (e.g., medical notes and corporate reports). Big data leads to big opportunities to uncover structures of real-world entities (e.g., `person`, `company`, `product`) and relations (e.g., `employee_of`, `manufacture`) from massive text corpora. Can machines automatically identify person, organization, and location entities in a news corpus and use them to summarize recent news events (Fig. 1.1)? Can we mine different relations between proteins, drugs, and diseases from massive and rapidly emerging life science literature? How would one represent factual structures hidden in a collection of medical reports to support answering precise queries or running data mining tasks?

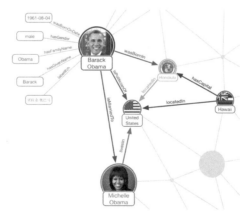

Figure 1.1: An illustration of entity and relation structures extracted from some text data. The nodes correspond to entities and the links represent their relationships.

While accessing documents in a gigantic collection is no longer a hard thing with the help of data management and information retrieval systems, people, especially those who are not domain experts, struggle to gain insights from such a large volume of text data: document understanding calls for in-depth content analysis, content analysis itself may require domain-specific knowledge, and over a large corpus, a complete read and analysis by domain experts will invariably be subjective, time-consuming, and relatively costly. Moreover, text data is highly diverse: Corpora from different domains, genres or languages typically require effective processing of a wide range of language resources (e.g., grammars, vocabularies, gazetteers). The "massive" and "messy" nature of text data poses significant challenges to creating tools for automated processing and algorithmic analysis of contents that scale with text volume.

This book introduces principled and scalable methods for the mining of typed entity and relation structures from unstructured text corpora, with a focus on overcoming the barriers in dealing with text corpora of various domains, genres, and languages. State-of-the-art information extraction (IE) approaches have relied on large amounts of task-specific labeled data (e.g., annotating terrorist attack-related entities in web forum posts written in Arabic), to construct machine-learning models (e.g., deep neural networks). However, even though domain experts can manually create high-quality training data for specific tasks as needed, both the *scale* and *efficiency* of such a manual process are limited. The research discussed in this book harnesses the power of "big text data" and focuses on creating generic solutions for *efficient construction of customized machine-learning models for factual structure extraction*, relying on only limited amounts of (or even no) task-specific training data.

The main coverage of this book includes: (1) *entity recognition and typing*, which automatically identifies token spans of real-world entities of interests from text and classifies them into a set of coarse-grained entity types; (2) *relation extraction*, which determine what kind of relations is expressed between two entities based on the sentences where they co-occur; and (3) *automated factual structure mining*, which aims to extract open-domain factual structures such as entity attributes and open-domain relation tuples. We provide scalable algorithmic approaches that leverage external knowledge bases (KBs) as sources of supervision and exploit data redundancy in massive text corpora, and we show how to use them in large-scale, real-world applications, including structured exploration and analysis of life sciences literature, extracting document facets from technical documents, document summarization, entity attribute discovery, and open-domain information extraction.

1.1 OVERVIEW OF THE BOOK

This book studies how to automate the process of extracting factual structures from a large corpus *with light human efforts* (i.e., Effort-Light StructMine), that is, with no task-specific manual annotation on the corpus. In contrast to existing knowledge base (KB) population approaches (e.g., Google's Knowledge Vault [Dong et al., 2014], NELL [Carlson et al., 2010], KnowItAll [Etzioni et al., 2004], DeepDive [Shin et al., 2015]) that harvest facts incrementally from the whole Web to cover common knowledge in the world, our approach aims to generate a structured (typed) view of all the entities and their relationships in a given corpus, to enable semantic, holistic, and fast analysis of all contents in the full corpus. Thus the extraction of a corpus-specific entity/relation structures is distinct from, but complements the task of KB population. As a result, the effort-light StructMine techniques for extracting entity and relation structures focus on establishing only corpus-specific factual knowledge (e.g., identifying the entities and relations disambiguated for that corpus), a task that is outside the scope of general KBs or graphs (see Fig. 1.2).

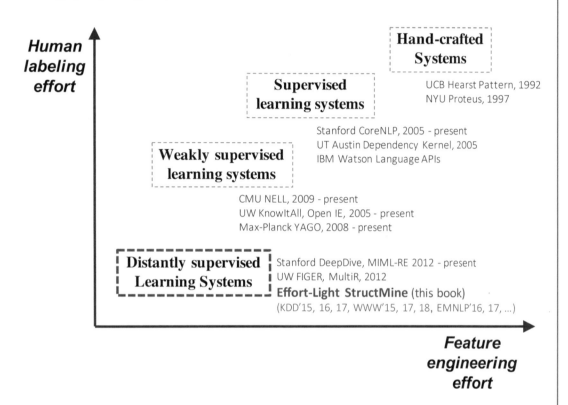

Figure 1.2: Overview of the related work. Our method, effort-light StructMine, has relied on lightest efforts on human laboring and feature engineering when compared with prior arts.

Challenges. We have witnessed the great success of machine-learning approaches in yielding state-of-the-art performance on information extraction when abundant amounts of training data are available. In contrast to rule-based systems, supervised learning-based systems shift the human expertise in customizing systems away from the complex handcrafting of extraction rules to the annotation of training data and feature engineering. The resulting effectiveness of supervised learning systems largely depends on the amount of available annotated training data and complexity of the task. When the quantity of annotated data is limited and the complexity of the task is high, these factors become bottlenecks in extracting corpus-specific entity/relation structures. Recent advances in bootstrapping pattern learning (e.g., NELL [Carlson et al., 2010], KnowItAll [Etzioni et al., 2004], OpenIE [Banko et al., 2007]) aim to reduce the amount of human involvement—only an initial set of annotated examples/patterns is required from domain experts, to iteratively produce more patterns and examples for the task. Such a process, however, still needs manual spot-checking of system intermediate output on a regular basis to avoid error propagation, and suffers from low coverage on "implicit relations," that is, those that are not overly expressed in the corpus and so fail to match textual patterns generated by the systems.

Framework of the Effort-Light Approach. Our solution to effort-light StructMine aims to bridge the gap between customized machine-learning models and the absence of high-quality task-specific training data. It leverages the information overlap between background facts stored in external KBs (e.g., Freebase [Bollacker et al., 2008], BioPortal [Noy et al., 2009]) and the given corpus to automatically generate large amounts of (possibly noisy) task-specific training data; and it exploits redundant text information within the massive corpus to reduce the complexity of feature generation (e.g., sentence parsing). This solution is based on two key intuitions outlined below.

First, in a massive corpus, structured information about some of the entities (e.g., entity types, relationships to other entities) can be found in external KBs. Can we align the corpus with external KBs to automatically generate training data for extracting entity and relation structures at a large scale? Such retrieved information supports the automated annotation of entities and relations in text and labeling of their categories, yielding (possibly noisy) corpus-specific training data (Figure 1.3). Although the overlaps between external KBs and the corpus at hand might involve only a small proportion of the corpus, the scale of the automatically labeled training data could still be much larger than that of manually annotated data by domain experts.

Second, text units (e.g., word, phrase) co-occur frequently with entities and other text units in a massive corpus. Can we exploit the textual co-occurrence patterns to characterize the semantics of text units, entities, and entity relationships? For example, having observed that *"government," "speech," "party"* co-occur frequently with `politician` entities in the training data, the next time these text units occur together with an unseen entity in a sentence, the algorithm can more confidently guess that entity is a `politician`. As such patterns become more apparent in a massive corpus with rich data redundancy, big text data leads to big opportunities in representing semantics of text unit without complex feature generation.

To systematically model the intuitions above, effort-light StructMine approaches the structure extraction tasks as follows: (1) annotate the text corpus automatically with target factual structure instances (e.g., entity names, entity categories, relationships) by referring to external KBs, to create a task-specific training data (i.e., distant supervision); (2) extract shallow text units (e.g., words, n-grams, word shapes) surrounding the annotated instances in local context; (3) learn semantic vector representations for target instances, text units, and their category labels based on distant supervision and corpus-level co-occurrence statistics, through solving joint optimization problems; and (4) apply learned semantic vector representations to extract new factual instances in the remaining part of the corpus. The resulting framework, which integrates these ideas, has minimal reliance on human efforts, and thus can be ported to solve structure extraction

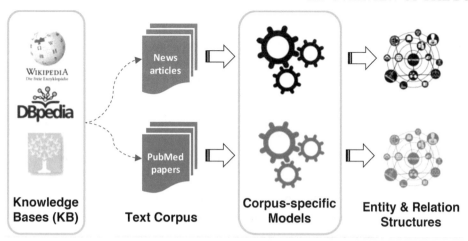

Figure 1.3: Illustration of the proposed framework. Effort-light StructMine leverages existing structures stored in external KBs to automatically generate large amounts of corpus-specific, potentially noisy training data, and builds corpus-specific models for extracting entity and relation structures.

tasks on text corpora of different kinds (i.e., **domain-independent**, **language-independent**, and **genre-independent**).

The book is organized into three main parts: (1) identifying typed entities, (2) extracting entity relationships, and (3) toward automated factual structure mining. Altogether, these steps form the pipeline to construct a structured knowledge network from a massive text corpus. We summarize the main problems of each part in the form of questions in Table 1.1.

1.1.1   PART I: IDENTIFYING TYPED ENTITIES

Real-world entities are important factual structures that can be identified from text to represent the factual knowledge embedded in massive amounts of text documents, and can serve as fundamental building blocks for many downstream data mining and natural language processing tasks such as KB construction and recommender systems. At a macroscopic level, how can we extract entities of types of interests from text with minimal reliance on labeled training data? At a microscopic level, how can we determine the fine-grained categories of an entity based on the context where it occurs? Our work proposes effective and distantly supervised methods for recognizing and typing entities from text by leveraging external KBs and exploiting rich data redundancy in large corpora. With the entities and their coarse-grained types extracted, we further look into how to assign more fine-grained entity types given the context and noisy distant supervision. Apart from entity recognition and typing, another important issue is resolving entity names that refer to the same real-world entities. How can computational systems

Table 1.1: Book organization

Part	Research Problem	Chapter
I: Identifying Typed Entities	**Entity Recognition and Typing:** How can we identify token spans of real-world entities and their types from text?	4
	Fine-grained Entity Typing: How can we assign fine-grained types to mentions of entities in text?	5
	Entity Synonym Discovery: How can we extract additional synonyms for known entities from large text corpora?	6
II: Extracting Typed Entity Relationships	**Joint Extraction of Entities and Relations:** What types of entities are mentioned in text and what types of relationships are expressed between them?	7
	Pattern-enhanced Embedding Learning for Relation Extraction: Can we combine distributional representation learning with pattern-based bootstrapping?	8
	Heterogeneous Supervision for Relation Extraction: Can we integrate distant supervision with domain expert rules for relation extraction in a principled way?	9
	Indirection Supervision for Relation Extraction: Can we leverage auxiliary question-answer pairs for relation extraction in a principled way?	10
III: Toward Automated StructMine	**Mining Entity Attribute Values:** Can we extract entity attributes and their values from massive text corpora without human supervision?	11
	Open Information Extraction: Can we extract quality entity-relationship tuples from text corpora in a holistic manner?	12
	Applications of Effort-Light StructMine	13

automatically identify synonyms of entities from massive text corpora or other kinds of data sources? Extracting a complete list of synonym strings for entities of interests can benefit many downstream applications. For example, when doing web search or information retrieval, one can leverage entity synonyms to enhance the process of query expansion. In topic modeling, by forcing synonyms of an entity to be assigned in the same latent topic, one can constrain the topic modeling process to yield high-quality topic representations. In this part of the book, we introduce methods that automate the process of identifying entity synonyms from text corpus with the distant supervision from external KBs.

Entity Recognition and Typing

How can we identify token spans of real-world entities and their categories from text?

One of the most important factual structures in text is entity. Recognizing entities from text and labeling their types (e.g., `person`, `location`) enables effectives structured analysis of unstructured text corpora (Chapter 4). Traditional named entity recognition (NER) systems are usually designed for several major types and general domains, and so require additional steps for adaptation to a new domain and new types. Our method, ClusType [Ren et al., 2015], aims at identifying typed entities of interests from text without task-specific human supervision. While most existing NER methods treat the problem as sequence tagging task and require significant amounts of manually labeled sentences (with typed entities), ClusType makes use of entity information stored in freely-available KBs to create large amounts of (yet potentially noisy) labeled data and infers types of other entities mentioned in text in a robust and efficient way (see Fig. 1.4).

Figure 1.4: An example for illustration of entity recognition and typing.

We formalize the entity recognition and typing task as a distantly supervised learning problem. The solution workflow is: (1) detect entity mentions from a corpus; (2) map candidate entity mentions to KB entities of target types; and (3) use those confidently mapped {mention, type} pairs as labeled data to infer the types of remaining candidate mentions. ClusType runs data-driven phrase mining to generate entity mention candidates and relation phrases (thus having no reliance on pre-trained name recognizer), and enforces the principle that relation phrases should be softly clustered when propagating type information between their argument entities. We formulate a joint optimization to integrate type propagation via relation phrases and clustering of relation phrases.

Highlights:

- *Problem*: We study the problem of distantly supervised entity recognition and typing in a domain-specific corpus, where only a corpus and a reference KB are given as input.

- *Methodology*: We introduce an efficient, domain-independent phrase segmentation algorithm for extracting entity mentions and relation phrases. Entity types can be estimated for entity mentions by solving the clustering-integrated type propagation.

- *Effectiveness on real-world corpora*: Our experiments on three datasets of different genres—news, reviews, and tweets—demonstrate that ClusType achieves significant improvements over the state-of-the-art.

Fine-grained Entity Typing

How can we assign fine-grained types to mentions of entities in text?

ClusType provides a data-driven way to identify typed entities from text with the help from external KBs. It is able to distinguish types of entities at a coarse-grained level (e.g., `location` vs. `organization`) based on the context surrounding the entity mention. However, many downstream applications will benefit if one can assign type to an entity mention from a much larger set of fine-grained types (for example, over 100) from a tree-structured hierarchy. Fine-grained entity typing allows one entity mention to have multiple types, which together constitute a type path in the given type hierarchy, depending on the local context. This task requires in-depth modeling of local context and thus is challenging for relation phrase-based method like ClusType (see Fig. 1.5).

How can we build models to automatically estimate the fine-grained type path for entity mention, without heavy reliance on human supervision? This is the problem we address in Chapter 5. When external KBs are available for generating fine-grained labels, a key issue with such distant supervision is that it assigns types to entity mentions in a *context-agnostic* way—one entity (e.g., *Barack Obama*) can have multiple entity types (e.g., `person`, `politician`, and `writer`) in KB, however, in a specific context only some of these types may describe the entity properly. While prior arts follow supervised learning methods to train typing models on noisy distant supervision, we propose noise-aware embedding approaches [Ren et al., 2016a,b] to *denoise* the set of labels given by distant supervision as we are learning the embeddings for text features. Such weakly supervised learning models the correlation between "true" labels and features in a more reliable way and thus produces more effective feature embeddings for prediction.

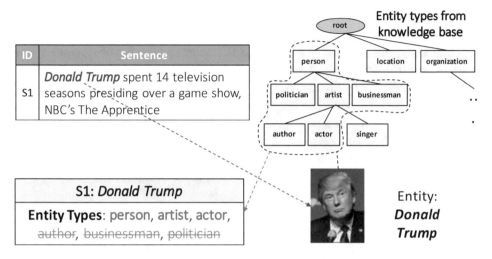

Figure 1.5: An illustration of the fine-grained entity typing problem.

Highlights:

- *Problem*: The first systematic study of noisy labels in distant supervision for entity typing problem.

- *Methodology 1*: An embedding-based framework, PLE [Ren et al., 2016a], is proposed that models and measures semantic similarity between text features and type labels and is robust to noisy labels.

- *Methodology 2*: A novel rank-based optimization problem is formulated to model noisy type label and type correlation [Ren et al., 2016b].

- *Effectiveness on real data*: The proposed methods achieve significant improvement over the state of the art on multiple fine-grained typing datasets, and demonstrate the effectiveness on recovering true labels from the noisy label set.

Synonym Discovery from Large Corpora

How can we extract additional synonyms for known entities from large text corpora?

In text corpora, people often use different strings to refer to the same entity. For example, in sport news about basketball, the journalist may start with talking about the basketball player *Kobe Bryant* using string "*Kobe Bryant*," then referring him again using the string "*Kobe*," "*Lakers 24*," or "*black memba*." This makes text corpora from a specific domain becomes a good data source for mining entity synonyms. However, discovering entity synonyms from domain-specific corpus is very challenging in the sense that: (1) one name string can refer to different entities (e.g., "*apple*"); and (2) it requires lots of manually labeled training data to build effective machine learning models. Our method, DPE [Qu et al., 2017], overcomes these two challenges by discovering synonyms from corpus with KBs (Chapter 6). The list of synonyms stored for each entity in multiple KBs serve as distant supervision to help determine important text features for the task, and the set of name strings for each entity also disambiguate their meanings. As the amounts of distant supervision is limited (due to the limited coverage in KBs), the DPE framework learns effective models by integrating two kinds of mutually complementing signals, that is, *distributional feature representations* based on corpus-level co-occurrence statistics and *textual patterns* based on local context. DPE jointly optimizes these two kinds of signals during the training stage so that one could provide "additional supervision" for the other.

Highlights:

- *Problem*: We study the problem of automatic synonym discovery with KBs, i.e., aiming to discover missing synonyms for entities by collecting training seeds from KBs.

- *Methodology*: We introduce a novel approach DPE, which naturally integrates the distributional representation learning approaches and the pattern-based approaches.

- *Effectiveness on real data*: We conduct extensive experiments on three real-world text corpora using different KBs and show that DPE can effectively discover new synonyms for KB entities.

1.1.2 PART II: EXTRACTING TYPED ENTITY RELATIONSHIPS

Our studies on entity structure mining (Part I) provide the basic building blocks of entity relationships, that is, the entity arguments mentioned in text. To further structure the text corpus the next step is to identify typed relationships between entities based on the local context in sentences (i.e., relation extraction). Identifying typed relationships is key to structuring content from text corpora for a wide range of downstream applications. For example, when an extraction system finds a "produce" relation between "company" and "product" entities in news articles, it supports answering questions like "*what products does company X produce?*" Once extracted, such relationship information is used in many ways (e.g., as primitives in KB population and question-answering systems). Traditional relation extraction systems rely on human-annotated corpora for training supervised learning models. Can we design a domain-independent relation extraction system that can apply to text corpora from different domains in the absence of human-annotated, domain training data? To address this question, we propose a distant-supervised relation extraction method in Chapter 7, which is able to reference existing relationship information stored in external KBs as a source of supervision and integrate the extraction models for both entities and relationships.

Joint Extraction of Entities and Relationships

What types of entities are mentioned in text and what typed of relationships are expressed between them?

With facts about entities, their types, and the relationships between them stored in external KBs, one can automatically generate large amounts of (potentially noisy) labeled training data for building entity recognition models (Part I) and relation extraction models. Prior arts focusing on this task have two major limitations: (1) the relation extraction process is partitioned into several subtasks and solved incrementally, which leads to errors propagating cascading down the pipeline; and (2) the noises brought in during the automatic label generation process are ignore and machine learning models are directly trained over the noisy labeled data (see Fig. 1.6).

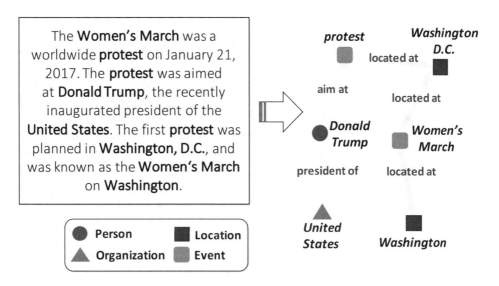

Figure 1.6: An illustration of the joint entity and relationship extraction problem.

To overcome these challenges, we study the problem of joint extraction of typed entities and relationships with KBs. Given a domain-specific corpus and an external KB, we aim to detect relation mentions together with their entity arguments from text, and categorize each in context by relation types of interests. Our method, CoType [Ren et al., 2017a], approaches the joint extraction task as follows: (1) it designs a domain-agnostic text segmentation algorithm to detect candidate entity mentions with distant supervision (i.e., minimal linguistic assumption); (2) it models the mutual constraints between the types of relation mentions and the types their entity arguments to enable feedbacks between the two subtasks; and (3) it models the true type labels in a candidate type set as latent variables and requires only the most confident type to be relevant to the mention. CoType achieves the state-of-the-art relation extraction performance under distant supervision, and demonstrates robust domain-independence across various datasets.

Highlights:

- *Methodology*: A novel distant supervision framework, CoType, is proposed, which extracts typed entities and relationships in domain-specific corpora with minimal linguistic assumption.

- *Effectiveness on real data*: Experiments with three public datasets demonstrate that CoType improves the performance of state-of-the-art systems of entity typing and relation extraction significantly, demonstrating robust domain-independence.

Pattern-enhanced Embedding Learning for Relation Extraction

Can we combine distributional representation learning with pattern-based bootstrapping?

Existing approaches to weakly supervised relation extraction can be divided into two kinds: the distributional methods and the pattern-based methods. Given a pair of entities, the distributional methods infer their relations based on the corpus-level statistics of both entities. Specifically, these methods try to learn low-dimensional representations of entities to preserve such statistics, so that entities with similar semantic meanings tend to have similar representations. Then a relation classifier can be learned using the given seed entity pairs, which takes entity representations as features for relation prediction. On the other hand, the pattern-based methods predict the relation between two entities from several sentences mentioning both of them. Toward this goal, traditional approaches try to extract discriminative textual patterns from such sentences, while recent approaches leverage deep neural networks for prediction. However, all these methods rely on a large number of seed entity pairs for training, and thus their performance is usually poor when there are only very limited seed entity pairs.

To overcome such difficulties, we introduce a co-training framework to integrate the distributional methods and the pattern-based methods, so that they can mutually provide extra supervision to overcome the scarcity problem of the seed entity pairs (Chapter 8). Specifically, the pattern module acts as a generator, as it can extract some candidate entity pairs based on the discovered patterns, whereas the distributional module is treated as a discriminator to evaluate the quality of each generated entity pair, that is, whether a pair has the target relation. To encourage the collaboration of both modules, we formulate a joint optimization process, in which we iterate between two sub-processes. During training, we keep iterating between the two sub-processes so that both methods can be consistently improved. Once the training converges, both methods can be applied to relation extraction, which extract new entity pairs from different perspectives.

Heterogeneous Supervision for Relation Extraction

Can we integrate distant supervision with domain expert rules for relation extraction in a principled way?

We introduce a general framework, *heterogeneous supervision* [Liu et al., 2017], which unifies various weak supervision sources for relation extraction (e.g., KB and domain-specific patterns). These supervisions often conflict with each other [Ratner et al., 2016]. To address these conflicts, prior work assumes that *a source is likely to provide true information with the same probability for all instances*. However, labeling functions, unlike human annotators, do not make casual mistakes but follow certain "error routine." Thus, the reliability of a labeling function is not consistent among different pieces of instances. In particular, a labeling function could be more reliable for a certain subset (also known as its *proficient subset*) compared to the rest. We identify these proficient subsets based on context information, only trust labeling functions on these subsets, and avoid assuming global source consistency.

This heterogeneous supervision framework captures context-sensitive semantic meaning through representation learning, and conducts both relation extraction and true label discovery in a context-aware manner. We embed relation mentions in a low-dimension vector space, where similar relation mentions tend to have similar relation types and annotations. True labels are further inferred based on reliabilities of labeling functions, which are calculated with their proficient subsets' representations. Then, these inferred true labels would serve as supervision for all components, including context representation, true label discovery and relation extraction. Besides, the context representation bridges relation extraction with true label discovery, and allows them to enhance each other.

Indirect Supervision for Relation Extraction

Can we leverage auxiliary question–answer pairs for relation extraction in a principled way?

Distant supervision (DS) replaces the manual training data generation with a pipeline that automatically links texts to a KB. However, the noise introduced to the automatically generated training data is not negligible. There are two major causes of error: incomplete KB and context-agnostic labeling process. If we treat unlinkable entity pairs as the pool of negative examples, false negatives can be commonly encountered as a result of the insufficiency of facts in KBs, where many true entity or relation mentions fail to be linked to KBs. On the other hand, context-agnostic labeling can engender false positive examples, due to the inaccuracy of the DS assumption that if a sentence contains any two entities holding a relation in the KB, the sentence must be expressing such relation between them.

Toward the goal of diminishing the negative effects of noisy DS training data, distantly supervised RE models that deal with training noise, as well as methods that directly improve the automatic training data generation process have been proposed. These methods mostly involve designing distinct assumptions to remove redundant training information. They do not have external trustworthy sources as guidance to uncover incorrectly labeled data. Without other separate information sources, the reliability of false label identification can be limited. Moreover, these noise reduction systems usually only address one type of error, either false positives or false negatives, although both types of error are observed to be significant. With the aim of overcoming the above two issues derived from relation extraction with distant supervision, we study the problem of relation extraction with indirect supervision from external sources. And due to the rapid emergence of QA systems as well as datasets of various QA tasks, we are motivated to leverage QA, a downstream application of relation extraction, to provide additional signals for learning RE models. We introduce a recently proposed method, ReQuest (Chapter 10). Given a domain-specific corpus and a set of target relation types from a KB, ReQuest detects relation mentions from the text and categorizes each in context by target types or Non-TargetType (None) by leveraging an independent dataset of QA pairs in the same domain.

1.1.3   PART III: TOWARD AUTOMATED FACTUAL STRUCTURE MINING

In this part of the book, we further explore how to go beyond pre-defined set of type labels for entities and their relationships, that is, open-domain factual structure mining. While methods discussed in prior chapters do not rely on task-specific human annotation data, they still require human experts to specify the set of types of interests for entities or relationships. In real-world applications, defining the type set is a challenging problem as the text corpus may cover multiple domains (thus the range and number of types are unknown), and there could be many different ways to define the type set.

Mining Entity Attribute Values with Meta Patterns

Can we extract entity attributes and their values from massive text corpora without human supervision?

Mining entity attributes and their values in news, tweets, papers, and many other kinds of text corpora has been an active theme in text mining and NLP research. Previous studies adopt a dependency parsing-based pattern discovery approach. However, the parsing results lose rich context around entities in the patterns, and the process is costly for a corpus of large scale. We introduce a novel *typed textual pattern structure*, called *meta pattern*, which is extended to a frequent, informative, and precise subsequence pattern in certain context. Based on meta patterns, we present an efficient framework, called MetaPAD (Chapter 11), which discovers meta patterns from massive corpora with three techniques: (1) it develops a context-aware segmentation method to carefully determine the boundaries of patterns with a learned pattern quality assessment function, which avoids costly dependency parsing and generates high-quality patterns; (2) it identifies and groups synonymous meta patterns from multiple facets—their types, contexts, and extractions; and (3) it examines type distributions of entities in the instances extracted by each group of patterns, and looks for appropriate type levels to make discovered patterns precise.

MetaPAD is designed to extract attributes of entities and the relations between entities by leveraging meta-patterns. It is a data-driven method and works for text corpora in multiple domains and languages. A meta-pattern consists of some abstractions of entities (e.g., using entity types to represent entities) and some free text (e.g., $PER, $DIGIT-year-old, etc.), which builds upon the results of phrase mining and entity typing. Its method essentially applies phrase mining (with more statistical features designed for this task) based on the abstracted text again. The major reason that it can extract more information is that the abstraction further increases data redundancy. Also, the discovery of synonym meta-patterns helps in the same manner.

Open Information Extraction with Global Structure Cohesiveness

Can we extract quality entity-relationship tuples from text corpora in a holistic manner?

Prior work on open-domain information extraction (Open IE) can be summarized as sharing two common characteristics: (1) conducting extraction only based on local context information and (2) adopting an incremental system pipeline. We introduce ReMine (Chapter 12), a novel open-domain information extraction approach that integrates local context signal and global structural signal in a unified framework with distant supervision.

Current Open IE systems focus on analyzing the local context within individual sentences to extract entity and their relationships, while ignoring the redundant information that can be collectively referenced across different sentences and documents in the corpus. Previously, ClusType has reduced local segmentation results via type propagation on the text co-occurrence graph constructed from corpus. With the same intuition, ReMine designs an effective way of measuring the quality of candidate relation tuples from rich information redundancy in a massive corpus. For example, seeing entity phrases "*London*" and "*Paris*" frequently co-occur with similar relation phrase and entities in the corpus, one gets to know that they have close semantics (same for "*Great Britain*" and "*France*"). On one hand, this helps confirm that (*Paris*, is in, *France*) is a quality tuple if knowing (*London*, is in , *Great Britain*) is a good tuple. On the other hand, this helps rule out the tuple (*Paris*, build, *new satellites*) as "*Louvre-Lens*" is semantically distant from "*Paris*" (i.e., rare to see "*Louvre-Lens*" and "*Paris*" co-occur with similar relation phrases or entities in the corpus). Therefore, the rich information redundancy in the massive corpus motivates us to design an effective way of measuring whether a candidate relation tuple is consistently used across various context in the corpus (i.e., global cohesiveness).

Most existing Open IE systems are composed of entity detection tools (e.g., named entity recognizer (NER) and noun phrase (NP) chunker) and relation tuple extraction module. The NERs and NP chunkers are often pre-trained for general domains and may not work well on a domain-specific corpus (e.g., biomedical papers or social media posts). Error propagation are inevitable in such a two-step pipeline. To address this problem, CoType adopts distant supervision to minimize language gap between different domains. Furthermore, CoType demonstrates that low-dimensional vector representations of entity and relations help reduce noise introduced by distant supervision. More generally, ReMine extracts general relation tuples rather than factual knowledge with specific relation types. Inspired by the successes of translation operation in modeling entity-relation interactions (e.g., e1 + r = e2 for the tuple (e1, r, e2)), ReMine measures the quality of extracted tuples via a translation-based objective function, and integrate it with a sentence segmentation objective for joint optimization. Overall, ReMine is an end-to-end pipeline that jointly optimizes both the extraction of entity and relation phrases and the global cohesiveness across the corpus. Global statistics are used to prune wrong extractions from local context, which leads to more valuable tuples. Experiments on massive real-world corpus demonstrate the effectiveness and robustness of ReMine when compared with other open IE systems.

Applications of Effort-Light Factual Structure Mining

In Chapter 13, we show that effort-light StructMine has been used in several downstream applications, led to a few lines of follow-up research and yielded real-world impact. We first discuss how to build on top of distant supervision to incorporate human supervision (e.g., curated rules from domain experts) in the effort-light StructMine framework, then show a real application on life sciences domain that makes use of the StructNet constructed by our methods, and introduce several other applications of our proposed work.

Building on top of our methods for extracting typed entities and relationships, we present a novel system, called Life-iNet, which transforms an *unstructured* background corpus into a *structured* network of factual knowledge, and supports multiple exploratory and analytic functions over the constructed network for knowledge discovery. To support analytic functionality, the Life-iNet system implements link prediction functions over the constructed network and integrates a distinctive summarization function to provide insightful analysis (e.g., answering questions such as "*which genes are distinctively related to the given disease type under GeneDiseaseAssociation relation?*").

The ClusType framework is general and can be applied to other kinds of classification tasks. In a recent work, this type of propagation method is extended to extract document facets. Given the large volume of technical documents available, it is crucial to automatically organize and categorize these documents to be able to understand and extract value from them. Toward this end, we introduce a new research problem called *facet extraction*. Given a collection of technical documents, the goal of facet extraction is to automatically label each document with a set of concepts for the key facets (e.g., application, technique, evaluation metrics, and dataset) that people may be interested in. Facet extraction has numerous applications, including document summarization, literature search, patent search and business intelligence.

In many use cases, people want to have a concise yet informative summary to describe the common and different places between two documents or two set of documents. One of our recent studies presents a novel research problem, *Comparative Document Analysis (CDA)*, that is, *joint* discovery of commonalities and differences between two individual documents (or two sets of documents) in a large text corpus. Given any pair of documents from a (background) document collection, CDA aims to automatically identify sets of quality phrases (entities) to summarize the commonalities of *both* documents and highlight the distinctions of each *with respect to the other* informatively and concisely.

Summary. The core of the book focuses on developing effective, human effort-light, and scalable methods for extracting factual structures from massive, domain-specific text corpora. Our contributions are in the area of text mining and information extraction, within which we focus on domain-independent and noise-robust approaches using distant supervision (in conjunction with publicly available KBs). The work has broad impact on a variety of applications: KB construction, question-answering systems, structured search and exploration of text data, recommender systems, network analysis, and many other text mining tasks. Next, we present research background on mining structured factual information and introduce related, useful notions and definitions.

CHAPTER 2

Background

In this chapter we introduce the key definitions and notions on information extraction and knowledge graph construction that are useful for understanding the methods and algorithms described in the book. At the end of this chapter we give a table with the common notations and their descriptions.

2.1 ENTITY STRUCTURES

We start with the definition of common text structures, followed by entity, and other entity-related concepts (e.g., entity mention, entity types).

Phrase: A phrase is a group of words (or possibly a single word) that functions as a constituent in the syntax of a sentence, a single unit within a grammatical hierarchy. Phrase acts as a semantic unit in a sentence with some special idiomatic meaning or other significance (e.g., "*machine learning*," "*watch TV*," "*before that happened*," "*too slowly*").

Noun Phrase: A noun phrase (or nominal phrase) is a phrase which has a noun as its head (i.e., the word that determines the syntactic category of that phrase). It usually consists of groups made up of nouns—a person, place, thing, or idea—and the modifiers such as determiners, adjectives, and conjunctions. When looking at the structure of language, we treat a noun phrase the same way we treat a common noun. Like all nouns, a noun phrase can be a subject, object, or complement in a sentence.

Example 2.1 Noun Phrase   In the sentence "The quick, brown fox jumped over the lazy dog," there are two noun phrases: "the quick, brown fox" and "the lazy dog." "the quick, brown fox" is the subject of the sentence and "the lazy dog" is the object.

Proper Name: A proper name is a noun phrase that in its primary application refers to a unique entity (e.g., *University of Southern California, Computer Science, United States*), as distinguished from a common noun which usually refers to a class of entities (e.g., *city, person, company*), or non-unique instances of a specific class (e.g., *this city, other people, our company*). When a noun refers to a unique entity, it is also called proper noun.

Entity: In information extraction and text mining, an *entity* (or *named entity*) is a real-world object, such as person, location, organization, product, and scientific concept, that can be *denoted* with a proper name. It can be abstract or have a physical existence. Examples of named entities include *Barack Obama*, *Chicago*, *University of Illinois*. An entity is denoted as *e* in this book.

Remark: Ambiguous Proper Names for Named Entity. In the expression of "named entity," the word *named* restricts the scope to those entities for which one or many strings stands consistently for some referent. In practice, one named entity may be referred by multiple proper names and one proper name may refer to multiple named entities. For example, the entity, *automotive company created by Henry Ford in 1903*, can be referred to proper names "*Ford*" or "*Ford Motor Company*," although "*Ford*" can refer to many other entities as well.

Entity Mention: An entity mention, denoted by *m*, is a token span (i.e., a sequence of words) in text that refers to a named entity. It consists of the proper name and the token index in the sentence.

Example 2.2   Entity Mention.   In the sentences "I had the **pulled pork sandwich** with **coleslaw** and **baked beans** for lunch. The **pulled pork sandwich** is the best I've tasted in **Phoenix!**," the entity mentions are bold-faced. The proper name "pulled pork sandwich" appears twice in the sentence, corresponding to the same named entity but different entity mentions (thus will have different entity mention IDs).

Entity Type: An entity type (or *entity class*, *entity category*) is a conceptual label for a collection of entities that share the same characteristics and attributes (e.g., `person`, `artist`, `singer`, `location`). Entities with the same entity types are similar to one another. Entity type instances refer to entities that are assigned with a specific entity type. In many applications, a set of entity types of interest are usually pre-specified by domain experts via providing example entity type instances. There also exist cases that entity types are related to each other (vs. mutually exclusive), forming a complex, DAG-structured type hierarchy.

Example 2.3   Entity Types in ACE Shared Task   The Automatic Content Extraction (ACE) Program [Doddington et al., 2004] was to develop information extraction technology to support automatic processing of natural language data. In the Entity Detection and Tracking (EDT) task of ACE, it focuses on seven types of entities: `Person`, `Organization`, `Location`, `Facility`, `Weapon`, `Vehicle`, and `Geo-Political Entity`. Each type was further divided into subtypes (for instance, `Organization` subtypes include `Government`, `Commercial`, `Education`, `Non-profit`, `Other`).

2.2   RELATION STRUCTURES

This section introduces the basic concepts on relations. We start with the definition of relation, followed by definitions of relation instance and mention.

Relation: a *relation* (or *relation type*, *relation class*), denoted as r, is a (pre-defined) predication about two or more entities. For example, from the sentence fragment *"Facebook co-found Mark Zuckerberg"* one can extract the FounderOf relation between entities *Mark Zuckerberg* and *Facebook*. In this book, we focus on binary relations, that is, relations between two entities.

Example 2.4   Relations in ACE Shared Task   Much of the prior work on extracting relations from text is based on the task definition from ACE program [Doddington et al., 2004]. A set of major relation types and their subtypes are defined by ACE. Examples of ACE major relation types include physical (an entity is physically near another entity), personal/social (a person is a family member of another person), and employment/affiliation (a person is employed by an organization).

Relation Instance: A relation instance denotes a relationship over two or more entities in a specific relation. When only considering binary relations, a relation instance can be represented as a triple with a pair of entities e_i and e_j, and their relation type r, i.e., (e_i, r, e_j).

Entity Argument: The two entities involved in a relation instance are referred to as entity arguments. The former is also referred to as *head entity* whereas the latter *tail entity*.

Relation Mention: A relation mention, z, denotes a specific occurrence of some relation instance in text. It records the two entity mentions for the pair of entity arguments, the relation type between these two entities, and the sentence s where the relation mention is found, i.e., $z = (m_i, r, m_j; s)$.

Example 2.5   Relation Mention   Suppose we are given two sentences: "Obama was born in Hawaii, USA" (s_1) and "Barack Obama, the president of United States" (s_2). There are two relation mentions between entities Barack Obama and United States: $z_1 = $ (Obama, BirthPlace, USA; s_1) and $z_2 = $ (Barack Obama, PresidentOf, United States; s_2). Although the entity arguments are the same, the two relation mentions have different relation types based on the sentence context.

2.3   DISTANT SUPERVISION FROM KNOWLEDGE BASES

KBs are emerging as a popular and useful way to represent and leverage codified knowledge for a variety of use cases. For example, codifying the key entities and relationships of a particular domain can greatly accelerate a variety of tasks from providing semantic and natural language search over more traditional business intelligence data, to providing enabling query expansion and matching, to discovery and exploration of related entities and relations extracted from a large corpus of unstructured documents.

In information extraction, the term "*knowledge base*" is most frequently used with large collections of curated, structured knowledge, such as WikiData [Vrandečić and Krötzsch, 2014], Freebase [Bollacker et al., 2008], YAGO [Suchanek et al., 2007], DBpedia [Auer et al., 2007] or CYC [Lenat, 1995]. It is also applied when such structured knowledge is automatically extracted, such as NELL [Carlson et al., 2010], SnowBall [Agichtein and Gravano, 2000], or OpenIE [Banko et al., 2007]. We use it to refer to any collection of relation triples, no matter their source or underlying ontology, if any. This can include subject-verb-object triples automatically extracted from large text corpora or curated from domain experts.

Fact: A fact in KB can refer to either a binary relation triple in the form of (e_i, r, e_j) or an is-A relation between an entity and a concept, such as *Facebook* is a company.

Formally, KB, Ψ, consists of a set of entities $\mathcal{E}_\Psi$ and curated facts on both relation instances $\mathcal{I}_\Psi$ and entity types $\mathcal{T}_\Psi$ (i.e., is-A relation between entities and their entity types). The set of relation types in the KB is denoted as $\mathbb{R}_\Psi$.

Example 2.6   Freebase   Curating a universal knowledge graph is an endeavor which is infeasible for most individuals and organizations. Thus, distributing that effort on as many shoulders as possible through crowdsourcing is a way taken by Freebase [Bollacker et al., 2008], a public, editable knowledge graph with schema templates for most kinds of possible entities (e.g., persons, cities, and movies). After MetaWeb, the company running Freebase, was acquired by Google, Freebase was shut down on March 31, 2015. The last version of Freebase contains roughly 50 million entities and 3 billion facts. Freebase's schema comprises roughly 27,000 entity types and 38,000 relation types.[1]

Example 2.7   DBpedia   DBpedia is a knowledge graph extracted from structured data in Wikipedia [Auer et al., 2007]. The main source for this extraction is the key-value pairs in the Wikipedia infoboxes. In a crowd-sourced process, types of infoboxes are mapped to the DBpedia ontology, and keys used in those infoboxes are mapped to properties in that ontology. Based on those mappings, a knowledge graph can be extracted. The most recent version of the main DBpedia (i.e., DBpedia 2016-10, extracted from the English Wikipedia based on dumps from October 2016) contains 6.6 million entities and 13 billion facts about those entities. The ontology comprises 735 classes and 2,800 relations.

---

[1]https://developers.google.com/freebase/

2.4   MINING ENTITY AND RELATION STRUCTURES

Here we describe the basic tasks in mining factual structures from text corpora, followed by a short introduction on related information extraction tasks.

Entity Recognition and Typing: Entity recognition and typing (or named entity recognition) addresses the problem of identification (detection) and classification of pre-defined types of entities, such as `organization` (e.g., "*United Nation*"), `person` (e.g., "*Barack Obama*"), `location` (e.g., "*Los Angeles*"), etc. The detection part aims to find the token span of entities mentioned in text (i.e., entity mention) and the classification part aims to assign the suitable type to entity mention based on its sentence context.

Fine-grained Entity Typing: The goal of fine-grained entity typing is to classify each entity mention m (based on its sentence context s) into a pre-defined set of types where the types are correlated and organized into a tree-structured type hierarchy $\mathcal{Y}$. Each entity mention will be assigned with an entity type path—a path in the given type hierarchy that may not end at a leaf node. For example, in Fig. 2.1, the entity mention "*Donald Trump*" is assigned with the type path "`person-artist-actor`" based on the given sentence.

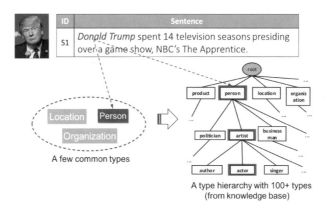

Figure 2.1: Illustration example for fine-grained entity typing.

Relation Extraction: Relation extraction is to detect and classify pre-defined relationships between entities recognized in text. In the corpus-level relation extraction setting, all sentences $\{s\}$ where a pair of entities (e_i, e_j) (proper names) occurs are collected as evidences for determining the appropriate relation type r. In the mention-level relation extraction, the correct label for a relation mention m is determined based on the sentence it occurs (i.e., s). In particular, a label (class) called "`None`" is included into the label set so as to classify a false positive candidate as "no relation."

2.5   COMMON NOTATIONS

We provide the most common notations and their brief definitions in Table 2.1. More specific notations used to explain proposed methods are introduced in the corresponding chapters.

Table 2.1: Common notations and definitions used throughout the book

Notation	Definition
s	Sentence
$d, \mathcal{D}$	Document, corpus
s	Entity
$t, \mathcal{T}$	Entity type, entity type set
m	Entity mention
$r, \mathbb{R}$	Relation type, relation type set
(e_i, r, e_j)	Relation instance of type r between entities e_i and e_j
z	Relation mention
$\mathcal{E}$	Finite set of entities in a corpus
$\mathcal{Z}$	Finite set of relationships between entities in a corpus
$G = (\mathcal{E}, \mathcal{Z})$	Directed, labeled graph that represents StructNet
Ψ	Knowledge base (e.g., Freebase, DBpedia)
$\mathcal{E}_\Psi$	Set of entities in KB
$\mathcal{T}_\Psi$	Entity types in KB
$\mathcal{I}_\Psi$	Set of relation instances in KB
$\mathbb{R}_\Psi$	Relation types in KB
$\mathcal{Y}$	Tree-structured entity type hierarchy

CHAPTER 3

Literature Review

This chapter provides an overview of prior arts and related studies on mining typed entities and relationships from text. Methods are categorized and organized based on the amounts of human labeled data required in the model training process, which also demonstrates the trajectory of research on reducing human supervision in entity and relation structure mining. We also review techniques developed for learning with noisy labeled data as well as open-domain information extraction, followed by a summary of our contributions.

The existing methods on mining entity and relation structures can be roughly categorized along two dimensions: (1) the amount of human supervision required and (2) the extraction task (problem formulation) that it is solving. Table 3.1 gives a few examples for each category. A method can be fully hand-crafted, supervised, weakly supervised, or distantly supervised. The second dimension is that the problem formulation of the task can be either sequence labeling (e.g., CRFs), transdutive classification (e.g., pattern bootstrapping, and label propagation), and inductive classification (e.g., SVM). More in-depth discussion about the literature related to concrete tasks and the proposed approaches can be found in each chapter.

3.1 HAND-CRAFTED METHODS

One straightforward way of finding entities and relationships in text is to write a set of textual patterns or rules (e.g., regular expression) for the different types of entities (or relationships between entities), with each pattern using some clue to identify the type of entity and relation. For example, in Fig. 3.1, the pattern "city such as NPList" is designed to extract a list of `city` entities from text. The NPList matches different noun phrases such as *New York*." Therefore, from the sentence in Fig. 3.1, this pattern is able to extract three entity mentions of `city` type, i.e., "*New York*," "*Los Angeles*," and "*Dallas*." Such patterns can be further enhanced with various lexical and syntactic constraints including part-of-speech tags and dependency parse structures.

Hand-crafted systems often rely on extensive lists of people, organizations, locations, and other entity or relation types. Many such lists are now available from the Web. These lists can be quite helpful, and for some entity types they are essential because good contextual patterns are not available, but they should be used with some caution. Systems which are primarily list based will suffer when new names arise, either because of changes in the input texts or the passage of time. Also, large lists may include entries which are proper names but more often are capitalized forms of common words. By building up a set of rules for each entity or relation type, it is possible to create a quite effective extraction system. However, obtaining high perfor-

Table 3.1: A few examples to give an idea about the categories of methods developed based on the amount of human supervision and the task

Method Category	Prior Work on Entity Extraction	Prior Work on Relation Extraction
Hand-crafted methods	DIPRE [Brin, 1998], FASTUS [Appelt, 1992]	Hearst Pattern [Hearst, 1992]
Supervised learning methods	Sequence tagging models like CRFs, HMMs, CMMs [Finkel et al., 2005; Ratinov and Roth, 2009]	Inductive classifiers like SVM, kernel methods [Bunescu and Mooney, 2005, Mooney and Bunescu, 2005], and various deep neural networks
Weakly-supervised methods	KnowItAll [Etzioni et al., 2005], SEISA [He and Xin, 2011], Gupta and Manning, 2014, semisupervised CRFs [Sarawagi, and Cohen, 2004]	SnowBall, NELL [Carlson et al., 2010]
Distantly-supervised methods	[Nakashole et al., 2013; Lin et al., 2012; Ling and Weld, 2012]	Mintz et al., 2009, MIMERE [Surdeanu et al., 2012; Riedel et al., 2010]
Open-domain extraction methods	Liberal IE [Huang et al., 2017], OpenIE systems [Banko et al., 2007; Fader et al., 2011; Schmitz et al, 2012, Angeli et al., 2015]	OpenIE systems [Banko et al., 2007; Fader et al., 2011; Schmitz et al, 2012, Angeli et al., 2015]

mance on the corpus of a specific domain, genre, or language does require some degree of skill. It also generally requires an annotated corpus which can be used to evaluate the rule set after each revision; without such a corpus there is a tendency—after a certain point—for added rules to actually worsen overall performance. It requires human experts to define rules or regular expressions or program snippets for performing the extraction. That person needs to be a domain expert and a programmer, and possess descent linguistic understanding to be able to develop robust extraction rules.

3.2   TRADITIONAL SUPERVISED LEARNING METHODS

If the creation of a high-quality entity and relation extractor by hand requires an annotated corpus, it is natural to ask whether extraction models can be trained automatically from such a

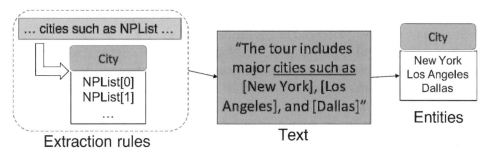

Figure 3.1: An illustration of hand-crafted extraction methods for extracting entity structures.

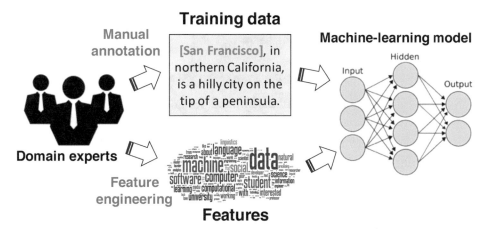

Figure 3.2: An illustration of supervised learning methods for extracting entity structures.

corpus. Such supervised methods for training a fully supervised extraction model will be considered in this section (see Fig. 3.2).

3.2.1   SEQUENCE LABELING METHODS

Many statistical learning-based named entity recognition algorithms treat the task as a sequence labeling problem. Sequence labeling is a general machine learning problem and has been used to model many natural language processing tasks including part-of-speech tagging, chunking, and named entity recognition. It can be formulated as follows. We are given a sequence of observations, denoted as $x = (x_1, x_2, \ldots, x_n)$. Usually each observation is represented as a feature vector. We would like to assign a label y_i to each observation x_i. While one may apply standard classification to predict the label y_i based solely on x_i, in sequence labeling, it is assumed that the label y_i depends not only on its corresponding observation x_i but also possibly on other observations and other labels in the sequence. Typically this dependency is limited to observations and labels within a close neighborhood of the current position i.

To map entity recognition to a sequence labeling problem, we treat each word in a sentence as an observation. The class labels have to clearly indicate both the boundaries and the types of named entities within the sequence. Usually the BIO notation, initially introduced for text chunking, is used. With this notation, for each entity type T, two labels are created, namely, $B-T$ and $I-T$. A token labeled with B-T is the beginning of a named entity of type T while a token labeled with $I-T$ is inside (but not the beginning of) a named entity of type T. In addition, there is a label O for tokens outside of any named entity.

Such supervised methods [Nadeau and Sekine, 2007, Ratinov and Roth, 2009] use fully annotated documents and different linguistic features to train sequence labeling model. To obtain an effective model, the amount of labeled data is significant [Ratinov and Roth, 2009], in spite of semi-supervised sequence labeling [Sarawagi and Cohen, 2004].

3.2.2   SUPERVISED RELATION EXTRACTION METHODS

Instead of creating and refining these patterns by hand, we can build a relation extractor from an annotated corpus. We will need a corpus which has been annotated both for entity mentions and for relations. We then want to convert relation tagging into a classification problem. To handle a single type of relation, we will train a classifier which classifies each pair of entity mentions appearing in the same sentence as either having or not having this relation. To handle n relation types, we can train an $(n + 1)$-way classifier. Alternatively, we can create two classifiers: a binary classifier which determines whether the entities bear some relation, and an n-way classifier (applied to instances which are passed by the first classifier) which determines which relation is involved. To apply this to new data, we first run an entity tagger and then apply the relation classifier to every pair of entities.

Traditional systems for relation extraction [Bach and Badaskar, 2007, Culotta and Sorensen, 2004, GuoDong et al., 2005] partition the process into several subtasks and solve them incrementally (i.e., detecting entities from text, labeling their types, and then extracting their relations). Such systems treat the subtasks independently and so may propagate errors across subtasks in the process. Recent studies [Li and Ji, 2014, Miwa and Sasaki, 2014, Roth and Yih, 2007] focus on joint extraction methods to capture the inherent linguistic dependencies between relations and entity arguments (e.g., the types of entity arguments help determine their relation type, and vice versa) to resolve error propagation.

3.3   WEAKLY SUPERVISED EXTRACTION METHODS

A weakly supervised method utilizes a small set of typed entities or relation instances as seeds, and extracts more entities or relationships of target types, which can largely reduce the amount of required labeled data.

3.3.1   SEMI-SUPERVISED LEARNING

A supervised method spares us the task of writing rules by hand but still requires substantial labor to prepare an annotated corpus. Can we reduce the amount of corpus that we need to annotate? This is possible through semi-supervised learning, including in particular those based on co-training [Blum and Mitchell, 1998, Nigam and Ghani, 2000]. A semi-supervised method makes use of limited amounts of labeled data together with large amounts of unlabeled data. The basic observation, already made use of in the long-range features of the supervised tagger, is that multiple instances of the same name have the same type. Some of these instances may occur in contexts which are indicative of a particular name type, and these may be used to tag other instances of the same name. In semi-supervised learning we apply these principles repeatedly: starting from a few common names of each type, we look for the contexts in which each of these names appears. If a context appears only with names of one type, we treat this as a predictive context; we look for other names which appear in this context and tag them with the predicted type. We add them to the set of names of known type, and the process repeats.

This is an example of co-training. In co-training, we have two views of the data—two sets of features which can predict the label on the data. In our case the two views are the context of a name and the "spelling" of a name (this includes the complete name and the individual tokens of the name). For each view, we are able to build a model from the labeled data; then we can use this model to label the unlabeled data and associate a confidence with each label. We build a model based on the first view, generate a label for each unlabeled datum, and keep the most confident labels, thus growing the labeled set. Then we do the same using the second view. Gradually the labeled set grows and the models are refined.

3.3.2   PATTERN-BASED BOOTSTRAPPING

Pattern-based bootstrapping [Gupta and Manning, 2014, Shi et al., 2010] derives patterns from contexts of seed entities or relation instances, and uses them to incrementally extract new entities/relationships and new patterns unrestricted by specific domains. However, pattern-based bootstrapping often suffers from low recall and semantic drift [Ling and Weld, 2012].

Iterative bootstrapping, such as probabilistic method [Nigam and Ghani, 2000] and label propagation [Talukdar and Pereira, 2010], softly assigns multiple types to an entity name and iteratively updates its type distribution, yet cannot decide the exact type for each entity mention based on its local context.

The DIPRE system [Brin, 1998] was one of the first systems for weakly supervised, pattern-based relation extraction, and one of the first designed to operate on the Web. The context patterns were based on character sequences before, between, and after the entities, and by doing so it could make use of both lexical contexts and XML mark-up contexts. The patterns were associated with particular web sites. There was no ranking of patterns or entity pairs; instead, some heuristics were used to ensure that the patterns were sufficiently specific. The entire procedure was demonstrated on the book–author relation. The Snowball system [Agichtein and

Gravano, 2000] introduced several improvements, including the use of name types, the ranking of patterns and entity pairs (as described above, using negative examples), and more relaxed patten matching. It was applied to the company–headquarters relation, a functional relation.

3.4   DISTANTLY SUPERVISED LEARNING METHODS

A distantly supervised method [Lin et al., 2012, Ling and Weld, 2012, Mintz et al., 2009, Nakashole et al., 2013, Riedel et al., 2010, Surdeanu et al., 2012] avoids expensive human labels by leveraging type information of entity and relation mentions which are confidently mapped to entries in KBs. Linked mentions are used to type those unlinkable ones in different ways, including training a contextual classifier [Nakashole et al., 2013], learning a sequence labeling model [Ling and Weld, 2012], and serving as labels in graph-based semi-supervised learning [Lin et al., 2012].

In the context of distant supervision, the label noise issue has been studied for other information extraction tasks such as relation extraction [Takamatsu et al., 2012]. In relation extraction, label noise is introduced by false positive textual matches of entity pairs. In entity typing, however, label noise comes from the assignment of types to entity mentions without considering their contexts. The forms of distant supervision are different in these two problems. Recently, Ren et al. [2016c] tackled the problem of label noise in fine-grained entity typing, but focused on how to generate a clean training set instead of doing entity typing.

3.5   LEARNING WITH NOISY LABELED DATA

Our proposed framework incorporates embedding techniques used in modeling words and phrases in large text corpora [Mikolov et al., 2013, Salehi et al., 2015, Yogatama et al., 2015], as well as nodes and links in graphs/networks [Perozzi et al., 2014, Tang et al., 2015]. These methods assume links are all correct (in unsupervised setting) or labels are all true (in supervised setting). CoType seeks to *model the true links and labels* in the embedding process (e.g., see our comparisons with LINE [Tang et al., 2015], DeepWalk [Perozzi et al., 2014], and FCM [Gormley et al., 2015]. Different from embedding structured KB entities and relations [Bordes et al., 2013, Toutanova et al., 2015], our task focuses on embedding entity and relation mentions in *unstructured* contexts.

In the context of modeling noisy labels, our work is related to partial-label learning [Cour et al., 2011, Nguyen and Caruana, 2008, Ren et al., 2016c] and multi-label multi-instance learning [Surdeanu et al., 2012], which deals with the problem where each training instance is associated with a set of noisy candidate labels (where *only one is correct*). Unlike these formulations, our *joint* extraction problem deals with both *classification with noisy labels* and *modeling of entity-relation interactions*. We also compare our full-fledged model with its variants CoType-EM and CoType-RM to validate the hypothesis on entity-relation interactions.

3.6   OPEN-DOMAIN INFORMATION EXTRACTION

Traditional techniques for mining entity and relation structures usually work on a corpus from a single domain (e.g., articles describing terrorism events), because the goal is to discover the most salient relations from such a domain-specific corpus. In some cases, however, our goal is to find all the potentially useful facts from a large and diverse corpus such as the Web. This is the focus of open information extraction, first introduced in Banko et al. [2007], then followed by several other open IE systems [Angeli et al., 2015, Fader et al., 2011, Schmitz et al., 2012].

Open information extraction does not assume any specific target relation type. It makes a single pass over the corpus and tries to extract as many relations as possible. Because no relation type is specified in advance, part of the extraction results is a phrase that describes the relation extracted. In other words, open information extraction generates (e_i, r, e_j) tuples where r is not from a finite set of pre-defined relation types, but can be arbitrary predicate or relation phrases. Banko et al. [2007] introduced an unlexicalized CRF-based method for open information extraction. The method is based on the observation that although different relation types have very different semantic meanings, there exist a small set of syntactic patterns that cover the majority of semantic relation mentions. It is therefore possible to train a relation extraction model that extracts arbitrary relations. The key is not to include lexical features in the model.

Extracting textual relations between subjective and objective from text has been extensively studied [Fader et al., 2011] and applied to entity typing [Lin et al., 2012]. Fader et al. [2011] utilize POS patterns to extract verb phrases between detected noun phrases to form relation assertion. Schmitz et al. [2012] further extend the textual relation by leveraging dependency tree patterns. These methods rely on linguistic parsers that may not generalize across domains. They also do not consider significance of the detected entity mentions in the corpus (see comparison with NNPLB [Lin et al., 2012]). There have been some studies on clustering and canonicalizing synonymous relations generated by open information extraction [Galárraga et al., 2014]. These methods either ignore entity type information when resolving relations, or assume that the types of relation arguments are already given.

PART I

Identifying Typed Entities

CHAPTER 4

Entity Recognition and Typing with Knowledge Bases

4.1 OVERVIEW AND MOTIVATION

Entity recognition is an important task in text analysis. Identifying token spans as entity mentions in documents and labeling their types (e.g., people, product or food) enables effective structured analysis of unstructured text corpus. The extracted entity information can be used in a variety of ways (e.g., to serve as primitives for information extraction [Schmitz et al., 2012] and KB population [Dong et al., 2014]). Traditional named entity recognition systems [Nadeau and Sekine, 2007, Ratinov and Roth, 2009] are usually designed for several major types (e.g., person, organization, location) and general domains (e.g., news), and so require additional steps for adaptation to a new domain and new types.

Entity-linking techniques [Shen et al., 2014] map from given entity mentions detected in text to entities in KBs like Freebase [Bollacker et al., 2008], where type information can be collected. But most of such information is manually curated, and thus the set of entities so obtained is of limited coverage and freshness (e.g., over 50% entities mentioned in Web documents are unlinkable [Lin et al., 2012]). The rapid emergence of large, domain-specific text corpora (e.g., product reviews) poses significant challenges to traditional entity recognition and entity-linking techniques and calls for methods of recognizing entity mentions of target types with minimal or no human supervision, and with no requirement that entities can be found in a KB.

There are broadly two kinds of efforts toward that goal: weak supervision and distant supervision. Weak supervision relies on manually specified seed entity names in applying pattern-based bootstrapping methods [Gupta and Manning, 2014, Huang and Riloff, 2010] or label propagation methods [Talukdar and Pereira, 2010] to identify more entities of each type. Both methods assume the seed entities are unambiguous and sufficiently frequent in the corpus, which requires careful seed entity selection by human [Kozareva and Hovy, 2010]. Distant supervision is a more recent trend, aiming to reduce expensive human labor by utilizing entity information in KBs [Lin et al., 2012, Nakashole et al., 2013] (see Fig. 4.1). The typical workflow is: (i) detect entity mentions from a corpus; (ii) map candidate mentions to KB entities of target types; and (iii) use those confidently mapped {mention, type} pairs as labeled data to infer the types of remaining candidate mentions.

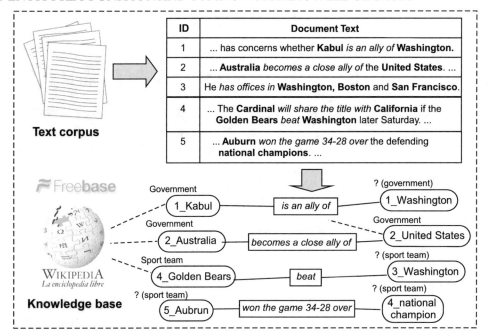

Figure 4.1: An example of distant supervision.

In this chapter, we study the problem of *distantly-supervised entity recognition in a domain-specific corpus*: Given a domain-specific corpus and a set of target entity types from a KB, we aim to effectively and efficiently detect entity mentions from that corpus, and categorize each by target types or Not-Of-Interest (NOI), with distant supervision. Existing distant supervision methods encounter the following limitations when handling a large, domain-specific corpus.

• **Domain Restriction:** They assume entity mentions are already extracted by existing entity detection tools such as noun phrase chunkers. These tools are usually trained on general-domain corpora like news articles (clean, grammatical) and make use of various linguistic features, but do not work well on specific, dynamic or emerging domains (e.g., tweets or restaurant reviews).

• **Name Ambiguity:** Entity names are often ambiguous—multiple entities may share the same surface name. In Fig. 4.1, for example, the surface name "*Washington*" can refer to either the U.S. government, a sport team, or the U.S. capital city. However, most existing studies [Huang and Riloff, 2010, Shen et al., 2012] simply output a type distribution for each surface name, instead of an exact type for *each* mention of the entity.

• **Context Sparsity:** Previous methods have difficulties in handling entity mentions with sparse context. They leverage a variety of contextual clues to find sources of shared semantics across different entities, including keywords [Talukdar and Pereira, 2010], Wikipedia concepts [Shen et al., 2012], linguistic patterns [Nakashole et al., 2013], and textual relations [Lin et al., 2012].

However, there are often many ways to describe even the same relation between two entities (e.g., "*beat*" and "*won the game 34-28 over*" in Fig. 4.1). This poses challenges on typing entity mentions when they are isolated from other entities or only share infrequent (sparse) context.

We address these challenges with several intuitive ideas. First, to address the domain restriction, we consider a domain-agnostic phrase mining algorithm to extract entity mention candidates with minimal dependence of linguistic assumption (e.g., part-of-speech (POS) tagging requires fewer assumptions of the linguistic characteristics of a domain than semantic parsing). Second, to address the name ambiguity, we do not simply merge the entity mention candidates with identical surface names but model each of them based on its surface name and contexts. Third, to address the context sparsity, we mine *relation phrases* co-occurring with the mention candidates, and infer synonymous relation phrases which share similar type signatures (i.e., express similar types of entities as arguments). This helps to form connecting bridges among entities that do not share identical context, but share synonymous relation phrases.

To systematically integrate these ideas, we develop a novel solution called **ClusType**. First, it mines both entity mention candidates and relation phrases by POS-constrained phrase segmentation; this demonstrates great cross-domain performance (Section 4.3.1). Second, it constructs a heterogeneous graph to faithfully represent candidate entity mentions, entity surface names, and relation phrases and their relationship types in a unified form (see Fig. 4.2). The entity mentions are kept as individual objects to be disambiguated, and linked to surface names and relation phrases (Sections 4.3.2–4.3.4). With the heterogeneous graph, we formulate a graph-based semi-supervised learning of two tasks jointly: (1) type propagation on graph and (2) relation phrase clustering. By clustering synonymous relation phrases, we can propagate types among entities bridged via these synonymous relation phrases. Conversely, derived entity argument types serve as good features for clustering relation phrases. These two tasks mutually enhance each other and lead to quality recognition of unlinkable entity mentions. In this chapter, we present an alternating minimization algorithm to efficiently solve the joint optimization problem, which iterates between type propagation and relation phrase clustering (Section 4.4). To our knowledge, this is the first work to *integrate entity recognition with textual relation clustering*.

The major novel contributions of this chapter are as follows: (1) we develop an efficient, domain-independent phrase mining algorithm for entity mention candidate and relation phrase extraction; (2) we propose a relation phrase-based entity recognition approach which models the type of each entity mention in a scalable way and softly clusters relation phrases, to resolve name ambiguity and context sparsity issues; (3) we formulate a joint optimization problem for clustering-integrated type propagation; and (4) our experiments on three datasets of different genres—news, Yelp reviews and tweets—demonstrate that the proposed method achieves significant improvement over the state-of-the-art (e.g., 58.3% enhancement in F1 on the Yelp dataset over the best competitor from existing work).

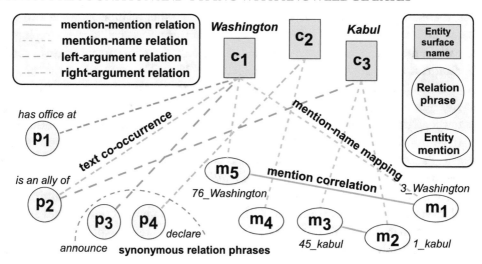

Figure 4.2: The constructed heterogeneous graph.

4.2 PROBLEM DEFINITION

The input to our proposed ER framework is a document collection $\mathcal{D}$, a KB Ψ with type schema $\mathcal{T}_\Psi$, and a *target type set* $\mathcal{T} \subset \mathcal{T}_\Psi$. In this work, we use the type schema of Freebase [Bollacker et al., 2008] and assume $\mathcal{T}$ is covered by Freebase.

An *entity mention*, m, is a token span in the text document which refers to a real-world entity e. Let c_m denote the *surface name* of m. In practice, people may use multiple surface names to refer to the same entity (e.g., "*black mamba*" and "*KB*" for *Kobe Bryant*). On the other hand, a surface name c could refer to different entities (e.g., "*Washington*" in Fig. 4.1). Moreover, even though an entity e can have multiple types (e.g., *J.F.K. airport* is both a location and an organization), the type of its specific mention m is usually unambiguous. We use a type indicator vector $\mathbf{t}y_m \in \{0,1\}^T$ to denote the entity type for each mention m, where $T = |\mathcal{T}| + 1$, i.e., m has type $t \in \mathcal{T}$ or is Not-of-Interest (NOI). By estimating $\mathbf{t}y_m$, one can predict type of m as type $(m) = \text{argmax}_{1 \leq i \leq T} y_{m,i}$.

Extracting textual relations from documents has been previously studied [Fader et al., 2011] and applied to entity typing [Lin et al., 2012, Nakashole et al., 2013]. A *relation phrase* is a phrase that denotes a unary or binary relation in a sentence [Fader et al., 2011] (see Fig. 4.3 for example). We leverage the rich semantics embedded in relation phrases to provide type cues for their entity arguments. Specifically, we define the *type signature* of a relation phrase p as two indicator vectors $\mathbf{t}p_L, \mathbf{t}p_R \in \mathbb{R}^T$. They measure how likely the left/right entity arguments of p belong to different types ($\mathcal{T}$ or NOI). A large positive value on $p_{L,t}$ ($p_{R,t}$) indicates that the left/right argument of p is likely of type t.

> Over:RP the weekend the system:EP dropped:RP nearly inches of snow in:RP
> western Oklahoma:EP and at:RP [Dallas Fort Worth International Airport]:EP sleet
> and ice caused:RP hundreds of [flight cancellations]:EP and delays. ...... It is
> forecast:RP to reach:RP [northern Georgia]:EP by:RP [Tuesday afternoon]:EP,
> Washington:EP and [New York]:EP by:RP [Wednesday afternoon]:EP,
> meteorologists:EP said:RP.

```
EP: entity mention candidate; RP: relation phrase
```

Figure 4.3: Example output of candidate generation.

Let $\mathcal{M} = \{m_1, \ldots, m_M\}$ denote the set of M candidate entity mentions extracted from $\mathcal{D}$. Suppose a subset of entity mentions $\mathcal{M}_L \subset \mathcal{M}$ can be confidently mapped to entities in Ψ. The type of a linked candidate $m \in \mathcal{M}_L$ can be obtained based on its mapping entity $\kappa_e(m)$ (see Section 4.4.1). This work focuses on predicting the types of *unlinkable candidate mentions* $\mathcal{M}_U = \mathcal{M} \setminus \mathcal{M}_L$, where $\mathcal{M}_U$ may consist of (1) mentions of the emerging entities which are not in Ψ; (2) new names of the existing entities in Ψ; and (3) invalid entity mentions. Formally, we define the problem of **distantly supervised entity recognition** as follows.

Problem 4.1   Entity Recognition and Typing. Given a document collection $\mathcal{D}$, a target type set $\mathcal{T}$, and a KB Ψ, our task aims to: (1) extract candidate entity mentions $\mathcal{M}$ from $\mathcal{D}$; (2) generate seed mentions $\mathcal{M}_L$ with Ψ; and (3) for each unlinkable candidate mention $m \in \mathcal{M}_U$, estimate its type indicator vector $\mathbf{t}y_m$ to predict its type. In our study, we assume each mention within a

sentence is only associated with a single type $t \in \mathcal{T}$. We also assume the target type set $\mathcal{T}$ is given (It is outside the scope of this study to generate $\mathcal{T}$). Finally, while our work is independent of entity linking techniques [Shen et al., 2014], our ER framework output may be useful to entity linking.

Framework Overview. Our overall framework is as follows.

1. Perform phrase mining on a POS-tagged corpus to extract candidate entity mentions and relation phrases, and construct a heterogeneous graph G to represent available information in a unified form, which encodes our insights on modeling the type for each entity mention (Section 4.3).

2. Collect seed entity mentions $\mathcal{M}_L$ as labels by linking extracted candidate mentions $\mathcal{M}$ to the KB Ψ (Section 4.4.1).

3. Estimate type indicator $\mathbf{t}y$ for unlinkable candidate mention $m \in \mathcal{M}_U$ with the proposed type propagation integrated with relation phrase clustering on G (Section 4.4).

4.3 RELATION PHRASE-BASED GRAPH CONSTRUCTION

We first introduce candidate generation in Section 4.3.1, which leads to three kinds of objects, namely candidate entity mentions $\mathcal{M}$, their surface names $\mathcal{C}$, and surrounding relation phrases $\mathcal{P}$. We then build a heterogeneous graph G, which consists of multiple types of objects and multiple types of links, to model their relationship. The basic idea for constructing the graph is that: the more two objects are likely to share the same label (i.e., $t \in \mathcal{T}$ or NOI), the larger the weight will be associated with their connecting edge.

Specifically, the constructed graph G unifies three types of links: *mention–name link* which represents the mapping between entity mentions and their surface names, *entity name–relation phrase link* which captures corpus-level co-occurrences between entity surface names and relation phrase, and *mention–mention link* which models distributional similarity between entity mentions. This leads to three subgraphs $G_{\mathcal{M},\mathcal{C}}$, $G_{\mathcal{C},\mathcal{P}}$, and $G_{\mathcal{M}}$, respectively. We introduce the construction of them in Sections 4.3.2–4.3.4.

4.3.1 CANDIDATE GENERATION

To ensure the extraction of informative and coherent entity mentions and relation phrases, we introduce a scalable, data-driven phrase mining method by incorporating both corpus-level statistics and syntactic constraints. Our method adopts a *global significance score* to guide the filtering of low-quality phrases and relies on a set of generic POS patterns to remove phrases with improper syntactic structure [Fader et al., 2011]. By extending the methodology used in El-Kishky et al. [2015], we can partition sentences in the corpus into non-overlapping segments which meet a significance threshold and satisfy our syntactic constraints. In doing so, entity candidates and relation phrases can be jointly extracted in an effective way.

First, we mine *frequent contiguous patterns* (i.e., sequences of tokens with no gap) up to a fixed length and aggregate their counts. A greedy agglomerative merging is then performed to form longer phrases while enforcing our syntactic constraints. Suppose the size of corpus $\mathcal{D}$ is N and the frequency of a phrase S is denoted by $v(S)$. The phrase-merging step selects the most significant merging, by comparing the frequency of a potential merging of two consecutive phrases, $v(S_1 \oplus S_2)$, to the expected frequency assuming independence, $N \frac{v(S_1)}{N} \frac{v(S_2)}{N}$. Additionally, we conduct syntactic constraint check on every potential merging by applying an entity check function $I_e(\cdot)$ and a relation check function $I_p(\cdot)$. $I_e(S)$ returns one if S is *consecutive nouns* and zero otherwise; and $I_p(S)$ return one if S (partially) matches one of the patterns in Table 4.2. Similar to Student's t-test, we define a score function $\rho_X(\cdot)$ to measure the significance and syntactic correctness of a merging [El-Kishky et al., 2015], where X can be e (entity mention) or p (relation phrase):

$$\rho_X(S_1, S_2) = \frac{v(S_1 \oplus S_2) - N \frac{v(S_1)}{N} \frac{v(S_2)}{N}}{\sqrt{v(S_1 \oplus S_2)}} \cdot I_X(S_1 \oplus S_2). \tag{4.1}$$

At each iteration, the greedy agglomerative algorithm performs the merging which has highest scores (ρ_e or ρ_p), and terminates when the next highest-score merging does not meet a pre-defined significance threshold. Relation phrases without matched POS patterns are discarded and their valid sub-phrases are recovered. Because the significance score can be considered analogous to hypothesis testing, one can use standard rule-of thumb values for the threshold (e.g., Z-score$\geq$2) [El-Kishky et al., 2015]. Overall, the threshold setting is not sensitive in our empirical studies. As all merged phrases are frequent, we have fast access to their aggregate counts and thus it is efficient to compute the score of a potential merging.

Figure 4.3 provides an example output of the candidate generation on *The New York Times* (NYT) corpus. We further compare our method with a popular noun phrase chunker[1] in terms of entity detection performance, using the extracted entity mentions. Table 4.1 summarizes the comparison results on three datasets from different domains (see Section 4.5 for details). Recall is most critical for this step, since we can recognize false positives in later stages of our framework, but no chance to later detect the misses, i.e., false negatives.

Table 4.1: Performance on entity detection

Method	NYT		Yelp		Tweet	
	Prec	Recall	Prec	Recall	Prec	Recall
Our method	**0.469**	**0.956**	**0.306**	**0.849**	0.226	**0.751**
NP chunker	0.220	0.609	0.296	0.247	**0.287**	0.181

Table 4.2: POS tag patterns for relation phrases

Pattern	Example
V	disperse; hit; struck; knock
P	in; at; of; from; to
V P	locate in; come from; talk to
VW*(P)	caused major damage on; come lately

V-verb; P-prep; W-{adv | adj | noun | det | pron}
W denotes multiple W; (P) denotes optional.

4.3.2   MENTION-NAME SUBGRAPH

In practice, directly modeling the type indicator for each candidate mention may be infeasible due to the large number of candidate mentions (e.g., $|\mathcal{M}| > 1$ million in our experiments). This

[1]TextBlob: http://textblob.readthedocs.org/en/dev/

results in an intractable size of parameter space, i.e., $\mathcal{O}(|M|T)$. Intuitively, both the entity name and the surrounding relation phrases provide strong cues on the type of a candidate entity mention. In Fig. 4.1, for example, the relation phrase "*beat*" suggests "*Golden Bears*" can mention a person or a sport team, while the surface name "*Golden Bears*" may refer to a sport team or a company. We propose to model the type indicator of a candidate mention based on the type indicator of its surface name and the type signatures of its associated relation phrases (see Section 4.4 for details). By doing so, we can reduce the size of the parameter space to $\mathcal{O}((|\mathcal{C}| + |\mathcal{P}|)T)$ where $|\mathcal{C}| + |\mathcal{P}| \ll |\mathcal{M}|$ (see Table 4.4 and Section 4.5.1). This enables our method to scale up.

Suppose there are n unique surface names $\mathcal{C} = \{c_1, \ldots, c_n\}$ in all the extracted candidate mentions $\mathcal{M}$. This leads to a biadjacency matrix $\Pi_{\mathcal{C}} \in \{0, 1\}^{M \times n}$ to represent the subgraph $G_{\mathcal{M},\mathcal{C}}$, where $\Pi_{\mathcal{C},ij} = 1$ if the surface name of m_j is c_j, and 0 otherwise. Each column of $\Pi_{\mathcal{C}}$ is normalized by its ℓ_2-norm to reduce the impact of popular entity names. We use a T-dimensional type indicator vector to measure how likely an entity name is subject to different types ($\mathcal{T}$ or NOI) and denote the type indicators for $\mathcal{C}$ by matrix $t_C \in \mathbb{R}^{n \times T}$. Similarly, we denote the type indicators for $\mathcal{M}$ by $t_Y \in \mathbb{R}^{M \times T}$.

4.3.3   NAME-RELATION PHRASE SUBGRAPH

By exploiting the aggregated co-occurrences between entity surface names and their surrounding relation phrases across multiple documents *collectively*, we weight the importance of different relation phrases for an entity name, and use their connected edge as bridges to propagate type information between different surface names by way of relation phrases. For each mention candidate, we assign it as the *left (right, resp.) argument* to the closest relation phrase appearing on its right (left, resp.) in a sentence. The *type signature* of a relation phrase refers to the two type indicators for its left and right arguments, respectively. The following hypothesis guides the type propagation between surface names and relation phrases.

Hypothesis 4.1: Entity-Relation Co-occurrences

If surface name c often appears as the left (right) argument of relation phrase p, then c's type indicator tends to be similar to the corresponding type indicator in p's type signature.

In Fig. 4.4, for example, if we know "*pizza*" refers to food and find it frequently co-occurs with the relation phrase "*serves up*" in its right argument position, then another surface name that appears in the right argument position of "*serves up*" is likely food. This reinforces the type propagation that "*cheese steak sandwich*" is also food.

Formally, suppose there are l different relation phrases $\mathcal{P} = \{p_1, \ldots, p_l\}$ extracted from the corpus. We use two biadjacency matrices $\Pi_L, \Pi_R \in \{0, 1\}^{M \times l}$ to represent the co-occurrences between relation phrases and their left and right entity arguments, respectively. We define $\Pi_{L,ij} = 1$ ($\Pi_{R,ij} = 1$) if m_i occurs as the *closest* entity mention on the left (right) of p_j *in a sentence*; and

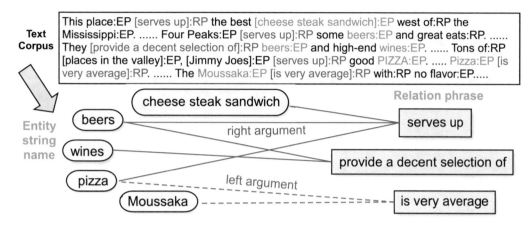

Figure 4.4: Example entity name-relation phrase links from Yelp reviews.

0 otherwise. Each column of Π_L and Π_R is normalized by its ℓ_2-norm to reduce the impact of popular relation phrases. Two bipartite subgraphs $G_{\mathcal{C},\mathcal{P}}$ can be further constructed to capture the aggregated co-occurrences between relation phrases $\mathcal{P}$ and entity names $\mathcal{C}$ across the corpus. We use two biadjacency matrices $tW_L, tW_R \in \mathbb{R}^{n \times l}$ to represent the edge weights for the two types of links, and normalize them:

$$tW_L = \Pi_{\mathcal{C}}^T \Pi_L \text{ and } tW_R = \Pi_{\mathcal{C}}^T \Pi_R;$$
$$tS_L = tD_L^{(\mathcal{C})-\frac{1}{2}} tW_L tD_L^{(\mathcal{P})-\frac{1}{2}} \text{ and } tS_R = tD_R^{(\mathcal{C})-\frac{1}{2}} tW_R tD_R^{(\mathcal{P})-\frac{1}{2}},$$

where tS_L and tS_R are normalized biadjacency matrices. For left-argument relationships, we define the diagonal surface name degree matrix $tD_L^{(\mathcal{C})} \in \mathbb{R}^{n \times n}$ as $D_{L,ii}^{(\mathcal{C})} = \sum_{j=1}^{l} W_{L,ij}$ and the relation phrase degree matrix $tD_L^{(\mathcal{P})} \in \mathbb{R}^{l \times l}$ as $D_{L,jj}^{(\mathcal{P})} = \sum_{i=1}^{n} W_{L,ij}$. Likewise, we define $tD_R^{(\mathcal{C})} \in \mathbb{R}^{n \times n}$ and $tD_R^{(\mathcal{P})} \in \mathbb{R}^{l \times l}$ based on tW_R for the right-argument relationships.

4.3.4   MENTION CORRELATION SUBGRAPH

An entity mention candidate may have an ambiguous name as well as associate with ambiguous relation phrases. For example, "*White House*" mentioned in the first sentence in Fig. 4.5 can refer to either an organization or a facility, while its relation phrase "*felt*" can have either a person or an organization entity as the left argument. It is observed that other co-occurring entity mentions (e.g., "*birth certificate*" and "*rose garden*" in Fig. 4.5) may provide good hints to the type of an entity mention candidate. We propose to propagate the type information between candidate mentions of *each* entity name based on the following hypothesis.

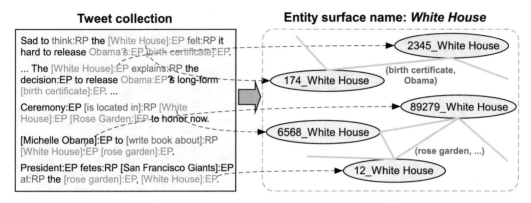

Figure 4.5: Example mention-mention links for entity surface name "*White House*" from Tweets.

Hypothesis 4.2: Mention correlation

If there exists a strong correlation (i.e., within sentence, common neighbor mentions) between two candidate mentions that share the same name, then their type indicators tend to be similar.

Specifically, for each candidate entity mention $m_i \in \mathcal{M}$, we extract the set of entity surface names which co-occur with m_i *in the same sentence*. An n-dimensional TF-IDF vector $\mathbf{f}^{(i)} \in \mathbb{R}^n$ is used to represent the importance of these co-occurring names for m_i where $f_j^{(i)} = v_s(c_j) \cdot \log(|\mathcal{D}|/v_{\mathcal{D}}(c_j))$ with term frequency in the sentence $v_s(c_j)$ and document frequency $v_{\mathcal{D}}(c_j)$ in $\mathcal{D}$. We use an affinity subgraph to represent the mention-mention link based on k-nearest neighbor (KNN) graph construction [He and Niyogi, 2004], denoted by an adjacency matrix $tW_{\mathcal{M}} \in \mathbb{R}^{M \times M}$. Each mention candidate is linked to its k most similar mention candidates which share the same name in terms of the vectors $\mathbf{f}$:.

$$W_{\mathcal{M},ij} = \begin{cases} \mathrm{sim}(\mathbf{f}^{(i)}, \mathbf{f}^{(j)}), & \text{if } \mathbf{f}^{(i)} \in N_k(\mathbf{f}^{(j)}) \text{ or } \mathbf{f}^{(j)} \in N_k(\mathbf{f}^{(i)}) \\ & \text{and } c(m_i) = c(m_j); \\ 0, & \text{otherwise,} \end{cases}$$

where we use the heat kernel function to measure similarity, i.e., $\mathrm{sim}(\mathbf{f}^{(i)}, \mathbf{f}^{(j)}) = \exp\left(-\|\mathbf{f}^{(i)} - \mathbf{f}^{(j)}\|^2/t\right)$ with $t = 5$ [He and Niyogi, 2004]. We use $N_k(\mathbf{f})$ to denote k nearest neighbors of $\mathbf{f}$ and $c(m)$ to denote the surface name of mention m. Similarly, we normalize $tW_{\mathcal{M}}$ into $tS_{\mathcal{M}} = tD_{\mathcal{M}}^{-\frac{1}{2}} tW_{\mathcal{M}} tD_{\mathcal{M}}^{-\frac{1}{2}}$ where the degree matrix $tD_{\mathcal{M}} \in \mathbb{R}^{M \times M}$ is defined by $D_{\mathcal{M},ii} = \sum_{j=1}^{M} W_{\mathcal{M},ij}$.

4.4   CLUSTERING-INTEGRATED TYPE PROPAGATION ON GRAPHS

This section introduces our unified framework for joint type propagation and relation phrase clustering on graphs.

A straightforward solution is to first perform hard clustering on the extracted relation phrases and then conduct type propagation between entity names and relation phrase clusters. Such a solution encounters several problems. One relation phrase may belong to multiple clusters, and the clusters so derived do not incorporate the type information of entity arguments. As such, the type prediction performance may not be best optimized by the mined clusters.

In our solution, we formulate a joint optimization problem to minimize both a graph-based semi-supervised learning error and a multi-view relation phrase clustering objective.

4.4.1 SEED MENTION GENERATION

We first collect type information for the extracted mention candidates $\mathcal{M}$ by linking them to the KB. This yields a set of type-labeled mentions $\mathcal{M}_L$. Our goal is then to type the remaining unlinkable mention candidates $\mathcal{M}_U = \mathcal{M}/\mathcal{M}_L$.

We utilize a state-of-the-art entity name disambiguation tool[2] to map each candidate mention to Freebase entities. Only the mention candidates which are mapped with high confidence scores (i.e., $\eta \geq 0.8$) are considered as valid output. We denote the mapping entity of a linked mention m as $\kappa_e(m)$, and the set of types of $\kappa_e(m)$ in Freebase as $\mathcal{T}(m)$. The linked mentions which associate with multiple target types (i.e., $|\mathcal{T}(m) \cap \mathcal{T}| > 1$) are discarded to avoid type ambiguity. This finally leads to a set of labeled (seed) mentions $\mathcal{M}_L$. In our experiments, we found that only a very limited amount of extracted candidate entity mentions can be confidently mapped to Freebase entities (i.e., $|\mathcal{M}_L|/|\mathcal{M}| < 7\%$). We define the type indicator $\mathbf{t}y_m$ for a linked mention $m \in \mathcal{M}_L$ as $\mathbf{t}y_{m,t} = 1$ if $\mathcal{T}(m) \cap \mathcal{T} = \{t\}$ and 0 otherwise, for $t \in \mathcal{T}$. Meanwhile, $\mathbf{t}y_{m,\text{NOI}}$ is assigned with 1 if $\mathcal{T}(m) \cap \mathcal{T} = \emptyset$ and 0 otherwise (Table 4.3).

Table 4.3: Statistics for seed generation

Dataset	NYT	Yelp	Tweet
# Extracted mentions	4.88 M	1.32 M	703 k
% Seed mentions	6.98	4.57	1.83
# Entities	17,326	5,662	12,211

4.4.2 RELATION PHRASE CLUSTERING

In practice, we observe that many extracted relation phrases have very few occurrences in the corpus. This makes it hard to model their type signature based on the aggregated co-occurrences with entity names (i.e., Hypothesis 4.1). In our experimental datasets, about 37% of the relation phrases have less than 3 unique entity surface names (in right or left arguments) in $G_{\mathcal{C},\mathcal{P}}$. Intuitively, by softly clustering synonymous relation phrases, the type signatures of frequent relation

---

[2]http://spotlight.dbpedia.org/

phrases can help infer the type signatures of infrequent (sparse) ones that have similar cluster memberships, based on the following hypothesis.

> **Hypothesis 4.3: Type signature consistency**
>
> If two relation phrases have similar cluster memberships, the type indicators of their left and right arguments (type signature) tend to be similar, respectively.

There has been some studies [Galárraga et al., 2014, Min et al., 2012] on clustering synonymous relation phrases based on different kinds of signals and clustering methods. We propose a general relation phrase clustering method to incorporate different features for clustering, which can be integrated with the graph-based type propagation in a mutually enhancing framework, based on the following hypothesis.

> **Hypothesis 4.4: Relation phrase similarity**
>
> Two relation phrases tend to have similar cluster memberships, if they have similar (1) strings, (2) context words, and (3) left and right argument type indicators.

In particular, type signatures of relation phrases have proven very useful in clustering of relation phrases which have infrequent or ambiguous strings and contexts [Galárraga et al., 2014]. In contrast to previous approaches, our method leverages the type information derived by the type propagation and thus does not rely strictly on external sources to determine the type information for all the entity arguments.

Formally, suppose there are n_s (n_c) unique words $\{w_1, \ldots, w_{n_s}\}$ ($\{w'_1, \ldots, w'_{n_c}\}$) in all the relation phrase strings (contexts). We represent the strings and contexts of the extracted relation phrases $\mathcal{P}$ by two feature matrices $\mathbf{f}_s \in \mathbb{R}^{l \times n_s}$ and $\mathbf{f}_c \in \mathbb{R}^{l \times n_c}$, respectively. We set $F_{s,ij} = 1$ if p_i contains the word w_j and 0 otherwise. We use a text window of 10 words to extract the context for a relation phrase from each sentence it appears in, and construct context features $\mathbf{f}_c$ based on TF-IDF weighting. Let $tP_L, tP_R \in \mathbb{R}^{l \times T}$ denote the type signatures of $\mathcal{P}$. Our solution uses the derived features (i.e., $\{\mathbf{f}_s, \mathbf{f}_c, tP_L, tP_R\}$) for multi-view clustering of relation phrases based on joint non-negative matrix factorization, which will be elaborated in the next section.

4.4.3   THE JOINT OPTIMIZATION PROBLEM

Our goal is to infer the label (type $t \in \mathcal{T}$ or NOI) for *each* unlinkable entity mention candidate $m \in \mathcal{M}_U$, i.e., estimating tY. We propose an optimization problem to unify two different tasks to achieve this goal: (i) type propagation over both the type indicators of entity names tC and the type signatures of relation phrases $\{tP_L, tP_R\}$ on the heterogeneous graph G by way of graph-based semi-supervised learning; and (ii) multi-view relation phrase clustering. The seed mentions $\mathcal{M}_L$ are used as initial labels for the type propagation. We formulate the objective function as follows:

$$\mathcal{O}_{\alpha,\gamma,\mu} = \mathcal{F}(\mathbf{t}C, \mathbf{t}P_L, \mathbf{t}P_R) + \mathcal{L}_\alpha(\mathbf{t}P_L, \mathbf{t}P_R, \{\mathbf{t}U^{(v)}, \mathbf{t}V^{(v)}\}, \mathbf{t}U^*)$$
$$+ \Omega_{\gamma,\mu}(\mathbf{t}Y, \mathbf{t}C, \mathbf{t}P_L, \mathbf{t}P_R). \tag{4.2}$$

The first term $\mathcal{F}$ follows from Hypothesis 4.1 to model *type propagation* between entity names and relation phrases. By extending local and global consistency idea [He and Niyogi, 2004], it ensures that the type indicator of an entity name is similar to the type indicator of the left (or right) argument of a relation phrase, if their corresponding association is strong:

$$\mathcal{F}(\mathbf{t}C, \mathbf{t}P_L, \mathbf{t}P_R) = \sum_{Z \in \{L,R\}} \sum_{i=1}^n \sum_{j=1}^l W_{Z,ij} \left\| \frac{\mathbf{t}C_i}{\sqrt{D_{Z,ii}^{(C)}}} - \frac{\mathbf{t}P_{Z,j}}{\sqrt{D_{Z,jj}^{(P)}}} \right\|_2^2. \tag{4.3}$$

The second term $\mathcal{L}_\alpha$ in Eq. (4.2) follows Hypotheses 4.3 and 4.4 to model the *multi-view relation phrase clustering* by joint non-negative matrix factorization. In this study, we consider each derived feature as one *view* in the clustering, i.e., $\{\mathbf{f}^{(0)}, \mathbf{f}^{(1)}, \mathbf{f}^{(2)}, \mathbf{f}^{(3)}\} = \{\mathbf{t}P_L, \mathbf{t}P_R, \mathbf{f}_s, \mathbf{f}_c\}$ and derive a four-view clustering objective as follows:

$$\mathcal{L}_\alpha(\mathbf{t}P_L, \mathbf{t}P_R, \{\mathbf{t}U^{(v)}, \mathbf{t}V^{(v)}\}, \mathbf{t}U^*) \tag{4.4}$$
$$= \sum_{v=0}^d \beta^{(v)} (\|\mathbf{f}^{(v)} - \mathbf{t}U^{(v)}\mathbf{t}V^{(v)T}\|_F^2 + \alpha\|\mathbf{t}U^{(v)}\mathbf{t}Q^{(v)} - \mathbf{t}U^*\|_F^2).$$

The first part of Eq. (4.4) performs matrix factorization on each feature matrix. Suppose there exists K relation phrase clusters. For each view v, we factorize the feature matrix $\mathbf{f}^{(v)}$ into a cluster membership matrix $\mathbf{t}U^{(v)} \in \mathbb{R}_{\geq 0}^{l \times K}$ for all relation phrases $\mathcal{P}$ and a type indicator matrix $\mathbf{t}V^{(v)} \in \mathbb{R}_{\geq 0}^{T \times K}$ for the K derived clusters. The second part of Eq. (4.4) enforces the consistency between the four derived cluster membership matrices through a *consensus matrix* $\mathbf{t}U^* \in \mathbb{R}_{\geq 0}^{l \times K}$, which applies Hypothesis 4.4 to incorporate multiple similarity measures to cluster relation phrases. As in Liu et al. [2013], we normalize $\{\mathbf{t}U^{(v)}\}$ to the same scale (i.e., $\|\mathbf{t}U^{(v)}\mathbf{t}Q^{(v)}\|_F \approx 1$) with the diagonal matrices $\{\mathbf{t}Q^{(v)}\}$, where $Q_{kk}^{(v)} = \sum_{i=1}^T V_{ik}^{(v)} / \|\mathbf{f}^{(v)}\|_F$, so that they are comparable under the same consensus matrix. A tuning parameter $\alpha \in [0, 1]$ is used to control the degree of consistency between the cluster membership of each view and the consensus matrix. $\{\beta^{(v)}\}$ are used to weight the information among different views, which will be automatically estimated. As the first part of Eq. (4.4) enforces $\{\mathbf{t}U^{(0)}, \mathbf{t}U^{(1)}\} \approx \mathbf{t}U^*$ and the second part of Eq. (4.4) imposes $\mathbf{t}P_L \approx \mathbf{t}U^{(0)}\mathbf{t}V^{(0)T}$ and $\mathbf{t}P_R \approx \mathbf{t}U^{(1)}\mathbf{t}V^{(1)T}$, it can be checked that $U_i^* \approx U_j^*$ implies both $P_{L,i} \approx P_{L,j}$ and $P_{R,i} \approx P_{R,j}$ for any two relation phrases, which captures Hypothesis 4.3.

The last term $\Omega_{\gamma,\mu}$ in Eq. (4.2) models the type indicator for each entity mention candidate, the mention-mention link and the supervision from seed mentions:

$$\Omega_{\gamma,\mu}(\mathbf{t}Y, \mathbf{t}C, \mathbf{t}P_L, \mathbf{t}P_R) = \|\mathbf{t}Y - f(\Pi_C \mathbf{t}C, \Pi_L \mathbf{t}P_L, \Pi_R \mathbf{t}P_R)\|_F^2$$
$$+ \frac{\gamma}{2} \sum_{i,j=1}^M W_{\mathcal{M},ij} \left\| \frac{\mathbf{t}Y_i}{\sqrt{D_{ii}^{(\mathcal{M})}}} - \frac{\mathbf{t}Y_j}{\sqrt{D_{jj}^{(\mathcal{M})}}} \right\|_2^2 + \mu\|\mathbf{t}Y - \mathbf{t}Y_0\|_F^2. \tag{4.5}$$

In the first part of Eq. (4.5), the type of each entity mention candidate is modeled by a function $f(\cdot)$ based on the the type indicator of its surface name as well as the type signatures of its associated relation phrases. Different functions can be used to combine the information from surface names and relation phrases. In this study, we use an equal-weight linear combination, i.e., $f(tX_1, tX_2, tX_3) = tX_1 + tX_2 + tX_3$. The second part follows Hypothesis 4.2 to model the mention-mention correlation by graph regularization, which ensures the consistency between the type indicators of two candidate mentions if they are highly correlated. The third part enforces the estimated tY to be similar to the initial labels from seed mentions, denoted by a matrix $tY_0 \in \mathbb{R}^{\mathcal{M} \times T}$ (see Section 4.4.1). Two tuning parameters $\gamma, \mu \in [0, 1]$ are used to control the degree of guidance from mention correlation in $G_{\mathcal{M}}$ and the degree of supervision from tY_0, respectively.

To derive the exact type of each candidate entity mention, we impose the 0-1 integer constraint $tY \in \{0, 1\}^{M \times T}$ and $tY1 = 1$. To model clustering, we further require the cluster membership matrices $\{tU^{(v)}\}$, the type indicator matrices of the derived clusters $\{tV^{(v)}\}$, and the consensus matrix tU^* to be non-negative. With the definition of $\mathcal{O}$, we define the joint optimization problem as follows:

$$\min_{\substack{tY, tC, tP_L, tP_R, tU^* \\ \{tU^{(v)}, tV^{(v)}, \beta^{(v)}\}}} \mathcal{O}_{\alpha, \gamma, \mu} \tag{4.6}$$

$$\text{s.t.} \quad \begin{aligned} &tY \in \{0, 1\}^{M \times T}, tY1 = 1, \ tU^* \geq 0, \\ &\{tU^{(v)}, \ tV^{(v)}\} \geq 0, \ \sum_{v=0}^{d} e^{-\beta^{(v)}} = 1, \end{aligned}$$

where $\sum_{v=0}^{d} e^{-\beta^{(v)}} = 1$ is used for avoiding trivial solution, i.e., solution which completely favors a certain view.

4.4.4   THE ClusType ALGORITHM

The optimization problem in Eq. (4.6) is mix-integer programming and thus is NP-hard. We propose a two-step approximate solution: first solve the real-valued relaxation of Eq. (4.6) which is a non-convex problem with $tY \in \mathbb{R}^{M \times T}$; then impose back the constraints to predict the exact type of each candidate mention $m_i \in \mathcal{M}_U$ by type $(m_i) = \text{argmax } Y_i$.

Directly solving the real-valued relaxation of Eq. (4.6) is not easy because it is non-convex. We develop an alternating minimization algorithm to optimize the problem with respect to each variable alternatively, which accomplishes two tasks iteratively: type propagation on the heterogeneous graph and multi-view clustering of relation phrases.

First, to learn the type indicators of candidate entity mentions, we take derivative on $\mathcal{O}$ with respect to tY while fixing other variables. As links only exist between entity mentions sharing the same surface name in $tW_{\mathcal{M}}$, we can efficiently estimate tY with respect to each entity name $c \in \mathcal{C}$. Let $tY^{(c)}$ and $tS_{\mathcal{M}}^{(c)}$ denote the sub-matrices of tY and $tS_{\mathcal{M}}$, which correspond to the candidate entity mentions with the name c, respectively. We have the update rule for $tY^{(c)}$

as follows:

$$tY^{(c)} = [(1 + \gamma + \mu)tI_c - \gamma t S_{\mathcal{M}}^{(c)}]^{-1}(\Theta^{(c)} + \mu tY_0^{(c)}), \quad \forall c \in \mathcal{C}, \tag{4.7}$$

where $\Theta = \Pi_C tC + \Pi_L tP_L + \Pi_R tP_R$. Similarly, we denote $\Theta^{(c)}$ and $tY_0^{(c)}$ as sub-matrices of Θ and tY_0 which correspond to the candidate mentions with name c, respectively. It can be shown that $[(1 + \gamma + \mu)tI_c - \gamma t S_{\mathcal{M}}^{(c)}]$ is positive definite given $\mu > 0$ and thus is invertible. Eq. (4.7) can be efficiently computed since the average number of mentions of an entity name is small (e.g., < 10 in our experiments). One can further parallelize this step to reduce the computational time.

Second, to learn the type indicators of entity names and the type signatures of relation phrases, we take derivative on $\mathcal{O}$ with respect to tC, tP_L, and tP_R while fixing other variables, leading to the following closed-form update rules:

$$tC = \frac{1}{2}[tS_L tP_L + tS_R tP_R + \Pi_C^T(tY - \Pi_L tP_L - \Pi_R tP_R)]; \tag{4.8}$$
$$tP_L = tX_0^{-1}[tS_L^T tC + \Pi_L^T(tY - \Pi_C tC - \Pi_R tP_R) + \beta^{(0)}tU^{(0)}tV^{(0)T}];$$
$$tP_R = tX_1^{-1}[tS_R^T tC + \Pi_R^T(tY - \Pi_C tC - \Pi_L tP_L) + \beta^{(1)}tU^{(1)}tV^{(1)T}];$$

where we define $tX_0 = [(1 + \beta^{(0)})tI_l + \Pi_L^T \Pi_L]$ and $tX_1 = [(1 + \beta^{(1)})tI_l + \Pi_R^T \Pi_R]$, respectively. Note that the matrix inversions in Eq. (4.8) can be efficiently calculated with linear complexity since both $\Pi_L^T \Pi_L$ and $\Pi_R^T \Pi_R$ are diagonal matrices.

Finally, to perform multi-view clustering, we first optimize Eq. (4.2) with respect to $\{tU^{(v)}, tV^{(v)}\}$ while fixing other variables, and then update tU^* and $\{\beta^{(v)}\}$ by fixing $\{tU^{(v)}, tV^{(v)}\}$ and other variables, which follows the procedure in Liu et al. [2013].

We first take the derivative of $\mathcal{O}$ with respect to $tV^{(v)}$ and apply Karush-Kuhn-Tucker complementary condition to impose the non-negativity constraint on it, leading to the multiplicative update rules as follows:

$$V_{jk}^{(v)} = V_{jk}^{(v)} \frac{[\mathbf{f}^{(v)T} tU^{(v)}]_{jk} + \alpha \sum_{i=1}^{l} U_{ik}^* U_{ik}^{(v)}}{\boldsymbol{\Delta}_{jk}^{(v)} + \alpha (\sum_{i=1}^{l} U_{ik}^{(v)2})(\sum_{i=1}^{T} V_{ik}^{(v)})}, \tag{4.9}$$

where we define the matrix $\boldsymbol{\Delta}^{(v)} = tV^{(v)} tU^{(v)T} tU^{(v)} + \mathbf{f}^{(v)-} tU^{(v)}$. It is easy to check that $\{tV^{(v)}\}$ remains non-negative after each update based on Eq. (4.9).

We then normalize the column vectors of $tV^{(v)}$ and $tU^{(v)}$ by $tV^{(v)} = tV^{(v)} tQ^{(v)-1}$ and $tU^{(v)} = tU^{(v)} tQ^{(v)}$. Following similar procedure for updating $tV^{(v)}$, the update rule for $tU^{(v)}$ can be derived as follows:

$$U_{ik}^{(v)} = U_{ik}^{(v)} \frac{[\mathbf{f}^{(v)+} tV^{(v)} + \alpha tU^*]_{ik}}{[tU^{(v)} tV^{(v)T} tV^{(v)} + \mathbf{f}^{(v)-} tV^{(v)} + \alpha tU^{(v)}]_{ik}}. \tag{4.10}$$

In particular, we make the decomposition $\mathbf{f}^{(v)} = \mathbf{f}^{(v)+} - \mathbf{f}^{(v)-}$, where $A_{ij}^+ = (|A_{ij}| + A_{ij})/2$ and $A_{ij}^- = (|A_{ij}| - A_{ij})/2$, in order to preserve the non-negativity of $\{tU^{(v)}\}$.

The proposed algorithm optimizes $\{tU^{(v)}, tV^{(v)}\}$ for each view v, by iterating between Eqs. (4.9) and (4.10) until the following reconstruction error converges:

$$\delta^{(v)} = \|\mathbf{f}^{(v)} - tU^{(v)}tV^{(v)T}\|_F^2 + \alpha\|tU^{(v)}tQ^{(v)} - tU^*\|_F^2. \tag{4.11}$$

With optimized $\{tU^{(v)}, tV^{(v)}\}$, we update tU^* and $\{\beta^{(v)}\}$ by taking the derivative on $\mathcal{O}$ with respect to each of them while fixing all other variables. This leads to the closed-form update rules as follows:

$$tU^* = \frac{\sum_{v=0}^{d} \beta^{(v)} tU^{(v)} tQ^{(v)}}{\sum_{v=0}^{d} \beta^{(v)}}; \quad \beta^{(v)} = -\log\left(\frac{\delta^{(v)}}{\sum_{i=0}^{d} \delta^{(i)}}\right). \tag{4.12}$$

Algorithm 4.1 summarizes our algorithm. For convergence analysis, ClusType applies block coordinate descent on the real-valued relaxation of Eq. (4.6). The proof procedure in Tseng [2001] (not included for lack of space) can be adopted to prove convergence for ClusType (to the local minimum).

Algorithm 4.1 The **ClusType** algorithm

Input: biadjacency matrices $\{\Pi_{\mathcal{C}}, \Pi_L, \Pi_R, tW_L, tW_R, tW_{\mathcal{M}}\}$, clustering features $\{\mathbf{f}_s, \mathbf{f}_c\}$, seed labels tY_0, number of clusters K, parameters $\{\alpha, \gamma, \mu\}$

1: Initialize $\{tY, tC, tP_L, tP_R\}$ with $\{tY_0, \Pi_{\mathcal{C}}^T tY_0, \Pi_L^T tY_0, \Pi_R^T tY_0\}$, $\{tU^{(v)}, tV^{(v)}, \beta^{(v)}\}$ and tU^* with positive values.
2: **repeat**
3:     Update candidate mention type indicator tY by Eq. (4.7)
4:     Update entity name type indicator tC and relation phrase type signature $\{tP_L, tP_R\}$ by Eq. (4.8)
5:     **for** $v = 0$ to 3 **do**
6:         **repeat**
7:             Update $tV^{(v)}$ with Eq. (4.9)
8:             Normalize $tU^{(v)} = tU^{(v)}tQ^{(v)}$, $tV^{(v)} = tV^{(v)}tQ^{(v)-1}$
9:             Update $tU^{(v)}$ by Eq. (4.10)
10:         **until** Eq. (4.11) converges
11:     **end for**
12:     Update consensus matrix tU^* and relative feature weights $\{\beta^{(v)}\}$ using Eq. (4.12)
13: **until** the objective $\mathcal{O}$ in Eq. (4.6) converges
14: **Predict** the type of $m_i \in \mathcal{M}_U$ by $\text{type}(m_i) = \text{argmax } Y_i$.

4.4.5 COMPUTATIONAL COMPLEXITY ANALYSIS

Given a corpus $\mathcal{D}$ with $N_{\mathcal{D}}$ words, the time complexity for our candidate generation and generation of $\{\Pi_{\mathcal{C}}, \Pi_L, \Pi_R, \mathbf{f}_s, \mathbf{f}_c\}$ is $\mathcal{O}(N_{\mathcal{D}})$. For construction of the heterogeneous graph G, the costs for computing $G_{\mathcal{C}, \mathcal{P}}$ and $G_{\mathcal{M}}$ are $\mathcal{O}(nl)$ and $\mathcal{O}(MM_{\mathcal{C}}d_{\mathcal{C}})$, respectively, where $M_{\mathcal{C}}$ denotes average number of mentions each name has and $d_{\mathcal{C}}$ denotes average size of feature dimensions ($M_{\mathcal{C}} < 10, d_{\mathcal{C}} < 5000$ in our experiments). It takes $\mathcal{O}(MT)$ and $\mathcal{O}(MM_{\mathcal{C}}^2 + l^2)$ time to initialize all the variables and pre-compute the constants in update rules, respectively.

We then study the computational complexity of ClusType in Algorithm 4.1 with the pre-computed matrices. In each iteration of the outer loop, ClusType costs $\mathcal{O}(MM_{\mathcal{C}}T)$ to update tY, $\mathcal{O}(nlT)$ to update tC and $\mathcal{O}(nT(K+l))$ to update $\{tP_L, tP_R\}$. The cost for inner loop is $\mathcal{O}(t_{in}lK(T+n_s+n_c))$ supposing it stops after t_{in} iterations ($t_{in} < 100$ in our experiments). Update of tU^* and $\{\beta^{(v)}\}$ takes $\mathcal{O}(lK)$ time. Overall, the computational complexity of ClusType is $\mathcal{O}(t_{out}nlT + t_{out}t_{in}lK(T+n_s+n_c))$, supposing that the outer loop stops in t_{out} iterations ($t_{out} < 10$ in our experiments).

4.5 EXPERIMENTS

4.5.1 DATA PREPARATION

Our experiments use three real-world datasets[3]: (1) **NYT:** constructed by crawling 2013 news articles from *The New York Times*. The dataset contains 118,664 articles (57M tokens and 480K unique words) covering various topics such as Politics, Business, and Sports; (2) **Yelp:** We collected 230,610 reviews (25M tokens and 418K unique words) from the 2014 *Yelp dataset challenge*; and (3) **Tweet:** We randomly selected 10,000 users in Twitter and crawled at most 100 tweets for each user in May 2011. This yields a collection of 302,875 tweets (4.2M tokens and 157K unique words).

1. Heterogeneous Graphs. We first performed lemmatization on the tokens using NLTK WordNet Lemmatizer[4] to reduce variant forms of words (e.g., eat, ate, eating) into their lemma form (e.g., eat), and then applied Stanford POS tagger [Toutanova et al., 2003] on the corpus. In candidate generation (see Section 4.3.1), we set maximal pattern length as 5, minimum support as 30, and significance threshold as 2, to extract candidate entity mentions and relation phrases from the corpus. We then followed the introduction in Section 4.3 to construct the heterogeneous graph for each dataset. We used the 5-nearest neighbor graphs when constructing the mention correlation subgraph. Table 4.4 summarizes the statistics of the constructed heterogeneous graphs for all three datasets.

Table 4.4: Statistics of the heterogeneous graphs

Dataset	NYT	Yelp	Tweet
# Entity mention candidates (M)	4.88M	1.32M	703K
# Entity surface names (n)	832K	195K	67K
# Relation phrases (l)	743K	271K	57K
# Links	29.32M	8.64M	3.59M
Average # mentions per string name	5.86	6.78	10.56

2. Clustering Feature Generation. Following the procedure introduced in Section 4.4.2, we used a text window of 10 words to extract the context features for each relation phrase (5 words

---

[3]Code and datasets used in this chapter can be downloaded at: `http://web.engr.illinois.edu/~xren7/clustype.zip`.
[4]`http://www.nltk.org/`

on the left and the right of a relation phrase), where stop-words are removed. We obtained 56K string terms (n_s) and 129K context terms (n_c) for the NYT dataset, 58K string terms and 37K context terms for the Yelp dataset, and 18K string terms and 38K context terms for the Tweet dataset, respectively, all unique term counts. Each row of the feature matrices was then normalized by its ℓ-2 norm.

3. Seed and Evaluation Sets. For evaluation purposes, we selected entity types which are popular in the dataset from Freebase, to construct the target type set $\mathcal{T}$. Table 4.5 shows the target types used in the three datasets. To generate the set of seed mentions $\mathcal{M}_L$, we followed the process introduced in Section 4.4.1 by setting the confidence score threshold as $\eta = 0.8$. To generate the evaluation sets, we randomly selected a subset of documents from each dataset and annotated them using the target type set $\mathcal{T}$ (each entity mention is tagged by one type). 1K documents are annotated for the NYT dataset (25,451 annotated mentions). 2.5K reviews are annotated for the Yelp dataset (21,252 annotated mentions). 3K tweets are annotated for the Tweet dataset (5,192 annotated mentions). We removed the mentions from the seed mention sets if they were in the evaluation sets.

Table 4.5: Target type sets $\mathcal{T}$ for the datasets

NYT	*person, organization, location, time_event*
Yelp	*food, time_event, job_title, location, organization*
Tweet	*time_event, business_consumer_product, person, location, organization, business_job_title, time_year_of_day*

4.5.2 EXPERIMENTAL SETTINGS

In our testing of ClusType and its variants, we set the number of clusters $K = \{4000, 1500, 300\}$ for NYT, Yelp, and Tweet datasets, respectively, based on the analyses in Section 4.5.3. We set $\{\alpha, \gamma, \mu\} = \{0.4, 0.7, 0.5\}$ by five-fold cross validation (of classification accuracy) on the seed mention sets. For convergence criterion, we stop the outer (inner) loop in Algorithm 4.1 if the relative change of $\mathcal{O}$ in Eq. (4.6) (reconstruction error in Eq. (4.11)) is smaller than 10^{-4}, respectively.

Compared Methods: We compared the proposed method (ClusType) with its variants which only model part of the proposed hypotheses. Several state-of-the-art entity recognition approaches were also implemented (or tested using their published codes): (1) **Stanford NER** [Finkel et al., 2005]: a CRF classifier trained on classic corpora for several major entity types; (2) **Pattern** [Gupta and Manning, 2014]: a state-of-the-art pattern-based bootstrapping method which uses the seed mention sets $\mathcal{M}_L$; (3) **SemTagger** [Huang and Riloff, 2010]: a bootstrapping method which trains contextual classifiers using the seed mention set $\mathcal{M}_L$ in a self-training manner; (4) **FIGER** [Ling and Weld, 2012]: FIGER trains sequence labeling models using automatically annotated Wikipedia corpora; (5) **NNPLB** [Lin et al., 2012]: It uses ReVerb assertions [Fader et al., 2011] to construct graphs and performs entity name-level label propagation; and (6) **APOLLO** [Shen et al., 2012]: APOLLO constructs heterogeneous graphs on entity mentions, Wikipedia concepts and KB entities, and then performs label propagation.

All compared methods were first tuned on our seed mention sets using five-fold cross validation. For ClusType, besides the proposed full-fledged model, **ClusType**, we compare (1) **ClusType-NoWm**: This variant does not consider mention correlation subgraph, i.e., set $\gamma = 0$ in ClusType; (2) **ClusType-NoClus**: It performs only type propagation on the heterogeneous graph, i.e., Eq. (4.4) is removed from $\mathcal{O}$; and (3) **ClusType-TwoStep**: It first conducts multi-view clustering to assign each relation phrase to a single cluster, and then performs ClusType-NoClus between entity names, candidate entity mentions and relation phrase clusters.

Evaluation Metrics: We use F1 score computed from Precision and Recall to evaluate the entity recognition performance. We denote the #system-recognized entity mentions as J and the # ground truth annotated mentions in the evaluation set as A. Precision is calculated by Prec $= \sum_{m \in J \cap A} \omega(t'_m = t_m)/|J|$ and Recall is calculated by Rec $= \sum_{m \in J \cap A} \omega(t'_m = t_m)/|A|$. Here, t_m and t'_m denote the true type and the predicted type for m, respectively. Function $\omega(\cdot)$ returns 1 if the predicted type is correct and 0 otherwise. Only mentions which have correct boundaries and predicted types are considered correct. For cross validation on the seed mention sets, we use classification accuracy to evaluate the performance.

4.5.3   EXPERIMENTS AND PERFORMANCE STUDY

1. Comparing ClusType with the other methods on entity recognition. Table 4.6 summarizes the comparison results on the three datasets. Overall, ClusType and its three variants outperform others on all metrics on NYT and Yelp and achieve superior Recall and F1 scores on Tweet. In particular, ClusType obtains a 46.08% improvement in F1 score and 168% improvement in Recall compared to the best baseline FIGER on the Tweet dataset and improves F1 by 48.94% compared to the best baseline, NNPLB, on the Yelp dataset.

FIGER utilizes a rich set of linguistic features to train sequence labeling models but suffers from low recall moving from a general domain (e.g., NYT) to a noisy new domain (e.g., Tweet) where feature generation is not guaranteed to work well (e.g., 65% drop in F1 score). Superior performance of ClusType demonstrates the effectiveness of our candidate generation and of the proposed hypotheses on type propagation over domain-specific corpora. NNPLB also utilizes textual relation for type propagation, but it does not consider entity surface name ambiguity. APOLLO propagates type information between entity mentions but encounters severe context sparsity issue when using Wikipedia concepts. ClusType obtains superior performance because it not only uses semantic-rich relation phrases as type cues for each entity mention, but also clusters the synonymous relation phrases to tackle the context sparsity issues.

Table 4.6: Performance comparisons on three datasets in terms of Precision, Recall, and F1 score

Datasets	NYT			Yelp			Tweet		
Method	Precision	Recall	F1	Precision	Recall	F1	Precision	Recall	F1
Pattern [Gupta and Manning, 2014]	0.4576	0.2247	0.3014	0.3790	0.1354	0.1996	0.2107	0.2368	0.2230
FIGER [Ling and Weld, 2012]	0.8668	0.8964	0.8814	0.5010	0.1237	0.1983	**0.7354**	0.1951	0.3084
SemTagger [Huang and Riloff, 2010]	0.8667	0.2658	0.4069	0.3769	0.2440	0.2963	0.4225	0.1632	0.2355
APOLLO [Shen et al, 2012]	0.9257	0.6972	0.7954	0.3534	0.2366	0.2834	0.1471	0.2635	0.1883
NNPLB [Lin et al., 2012]	0.7487	0.5538	0.6367	0.4248	0.6397	0.5106	0.3327	0.1951	0.2459
ClusType-NoClus	0.9130	0.8685	0.8902	0.7629	0.7581	0.7605	0.3466	0.4920	0.4067
ClusType-NoWm	0.9244	0.9015	0.9128	0.7812	0.7634	0.7722	0.3539	**0.5434**	0.4286
ClusType-TwoStep	0.9257	0.9033	0.9143	0.8025	0.7629	0.7821	0.3748	0.5230	0.4367
ClusType	**0.9550**	**0.9243**	**0.9394**	**0.8333**	**0.7849**	**0.8084**	0.3956	0.5230	**0.4505**

2. Comparing ClusType with its variants. Comparing with ClusType-NoClus and ClusType-TwoStep, ClusType gains performance from integrating relation phrase clustering with type propagation in a mutually enhancing way. It always outperforms ClusType-NoWm on Precision and F1 on all three datasets. The enhancement mainly comes from modeling the mention correlation links, which helps disambiguate entity mentions sharing the same surface names.

3. Comparing on different entity types. Figure 4.6 shows the performance on different types on Yelp and Tweet. ClusType outperforms all the others on each type. It obtains larger gain on *organization* and *person*, which have more entities with ambiguous surface names. This indicates that modeling types on entity mention level is critical for name disambiguation. Superior performance on *product* and *food* mainly comes from the domain independence of our method because both NNPLB and SemTagger require sophisticated linguistic feature generation which is hard to adapt to new types.

4. Comparing with trained NER. Table 4.7 compares ours with a traditional NER method, *Stanford NER*, trained using classic corpora like ACE corpus, on three major types—person, location, and organization. ClusType and its variants outperform Stanford NER on the corpora which are dynamic (e.g., NYT) or domain-specific (e.g., Yelp). On the Tweet dataset, ClusType has lower Precision but achieves a 63.59% improvement in Recall and 7.62% improvement in F1

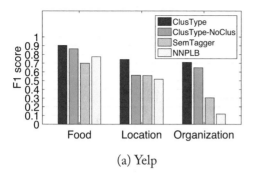
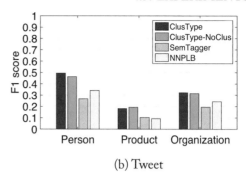

(a) Yelp   (b) Tweet

Figure 4.6: Performance breakdown by types.

Table 4.7: F1 score comparison with trained NER

Method	NYT	Yelp	Tweet
Stanford NER [Finkel et al., 2005]	0.6819	0.2403	0.4383
ClusType-NoClus	0.9031	0.4522	0.4167
ClusType	**0.9419**	**0.5943**	**0.4717**

score. The superior Recall of ClusType mainly comes from the domain-independent candidate generation.

5. Testing on sensitivity over the number of relation phrase clusters, K. Figure 4.7a, ClusType was less sensitive to K compared with its variants. We found on the Tweet dataset, ClusType achieved the best performance when $K = 300$ while its variants peaked at $K = 500$, which indicates that better performance can be achieved with fewer clusters if type propagation is integrated with clustering in a mutually enhancing way. On the NYT and the Yelp datasets (not shown here), ClusType peaked at $K = 4000$ and $K = 1500$, respectively.

6. Testing on the size of seed mention set. Seed mentions are used as labels (distant supervision) for typing other mentions. By randomly selecting a subset of seed mentions as labeled data (sampling ratio from 0.1–1.0), Fig. 4.7b shows ClusType and its variants are not very sensitive to the size of seed mention set. Interestingly, using all the seed mentions does not lead to the best performance, likely caused by the type ambiguity among the mentions.

7. Testing on the effect of corpus size. Experimenting on the same parameters for candidate generation and graph construction, Fig. 4.7c shows the performance trend when varying the sampling ratio (subset of documents randomly sampled to form the input corpus). ClusType and its variants are not very sensitive to the changes of corpus size, but NNPLB had over 17%

drop in F1 score when sampling ratio changed from 1.0–0.1 (while ClusType had only 5.5%). In particular, they always outperform FIGER, which uses a trained classifier and thus does not depend on corpus size.

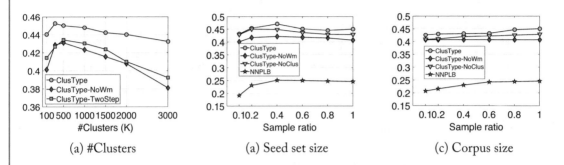

| (a) #Clusters | (a) Seed set size | (c) Corpus size |

Figure 4.7: Performance changes in F1 score with #clusters, #seeds, and corpus size on Tweets.

4.6   DISCUSSION

1. Example output on two Yelp reviews. Table 4.8 shows the output of ClusType, SemTagger, and NNPLB on two Yelp reviews: ClusType extracts more entity mention candidates (e.g., "*BBQ,*" "*ihop*") and predicts their types with better accuracy (e.g., "*baked beans,*" "*pulled pork sandwich*").

Table 4.8: Example output of ClusType and the compared methods on the Yelp dataset

ClusType	SemTagger	NNPLB
The best **BBQ:Food** I've tasted in **Phoenix:LOC** ! I had the **[pulled pork sandwich]:Food** with **coleslaw:-Food** and **[baked beans]:-Food** for lunch. ...	The best BBQ I've tasted in **Phoenix:LOC** ! I had the pulled [pork sandwich]:LOC with **coleslaw:Food** and [baked beans]:LOC for lunch. ...	The best BBQ:Loc I've tasted in **Phoenix:LOC** ! I had the pulled pork sandwich:Food with coleslaw and baked beans:Food for lunch:Food. ...
I only go to **ihop:LOC** for **pancakes:Food** because I don't really like anything else on the menu. Ordered **[chocolate chip pancakes]:Food** and a **[hot chocolate]:Food**.	I only go to ihop for pancakes because I don't really like anything else on the menu. Ordered [chocolate chip pancakes]:LOC and a [hot chocolate]:LOC.	I only go to ihop for pancakes because I don't really like anything else on the menu. Ordered chocolate chip pancakes and a hot chocolate.

2. Testing on context sparsity. The type indicator of each entity mention candidate is modeled in ClusType based on the type indicator of its surface name and the type signatures of its co-occurring relation phrases. To test the handling of different relation phrase sparsity, two groups of 500 mentions are selected from Yelp: mentions in *Group A* co-occur with frequent relation phrases (~4.6K occurrences in the corpus) and those in *Group B* co-occur with sparse relation phrases (~3.4K occurrences in the corpus). Figure 4.8a compares their F1 scores on the Tweet dataset. In general, all methods obtained better performance when mentions co-occurring with frequent relation phrases than with sparse relation phrases. In particular, we found that ClusType and its variants had comparable performance in Group A but ClusType obtained superior performance in Group B. Also, ClusType-TwoStep obtained larger performance gain over ClusType-NoClus in Group B. This indicates that clustering relation phrases is critical for performance enhancement when dealing with sparse relation phrases, as expected.

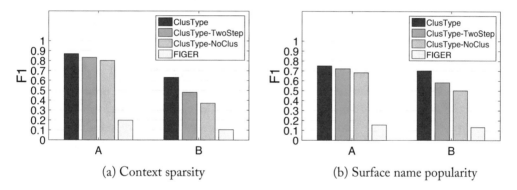

(a) Context sparsity   (b) Surface name popularity

Figure 4.8: Case studies on context sparsity and surface name popularity on the Tweet dataset.

3. Testing on surface name popularity. We generated the mentions in Group A with high frequency surface names (~2.7K occurrences) and those in *Group B* with infrequent surface names (~1.5). Figure 4.8b shows the degraded performance of all methods in both cases—likely due to ambiguity in popular mentions and sparsity in infrequent mentions. ClusType outperforms its variants in Group B, showing it handles well mentions with insufficient corpus statistics.

4. Example relation phrase clusters. Table 4.9 shows relation phrases along with their corpus frequency from three example relation phrase clusters for the NYT dataset ($K = 4000$). We found that not only synonymous relation phrases, but also both sparse and frequent relation phrases can be clustered together effectively (e.g., "*want hire by*" and "*recruited by*"). This shows that ClusType can boost sparse relation phrases with type information from the frequent relation phrases with similar group memberships.

Table 4.9: Example relation phrase clusters and their corpus frequency from the NYT dataset

ID	Relation Phrase
1	Recruited by (5.1K); employed by (3.4K); want hire by (264)
2	Go against (2.4K); struggling so much against (54); run for re-election against (112); campaigned against (1.3K)
3	Looking at ways around (105); pitched around (1.9K); echo around (844); present at (5.5K)

4.7   SUMMARY

Entity recognition is an important but challenging research problem. In reality, many text collections are from specific, dynamic, or emerging domains, which poses significant new challenges for entity recognition with increase in name ambiguity and context sparsity, requiring entity detection without domain restriction. In this work, we investigate entity recognition (ER) with distant-supervision and propose a novel relation phrase-based ER framework, called **ClusType**, that runs *data-driven* phrase mining to generate entity mention candidates and relation phrases, and enforces the principle that relation phrases should be *softly* clustered when propagating type information between their argument entities. Then we predict the type of *each* entity mention based on the type signatures of its co-occurring relation phrases and the type indicators of its surface name, as computed over the corpus. Specifically, we formulate a joint optimization problem for two tasks, *type propagation with relation phrases* and *multi-view relation phrase clustering*. Our experiments on multiple genres—news, Yelp reviews, and tweets—demonstrate the effectiveness and robustness of ClusType, with an average of 37% improvement in F1 score over the best compared method.

CHAPTER 5

Fine-Grained Entity Typing with Knowledge Bases

5.1 OVERVIEW AND MOTIVATION

Assigning types (e.g., person, organization) to mentions of entities in context is an important task in natural language processing (NLP). The extracted entity type information can serve as primitives for relation extraction [Mintz et al., 2009] and event extraction [Ji and Grishman, 2008], and assists a wide range of downstream applications including KB completion [Dong et al., 2014], question answering [Lin et al., 2012], and entity recommendation [Yu et al., 2014]. While traditional named entity recognition systems [Nadeau and Sekine, 2007, Ratinov and Roth, 2009] focus on a small set of coarse types (typically fewer than 10), recent studies [Ling and Weld, 2012, Yosef et al., 2012] work on a much larger set of fine-grained types (usually over 100) which form a tree-structured hierarchy (see the blue region of Fig. 5.1). Fine-grained typing allows one mention to have multiple types, which together constitute a *type-path* (not necessarily ending in a leaf node) in the given type hierarchy, *depending on the local context* (e.g., sentence). Consider the example in Fig. 5.1, *"Arnold Schwarzenegger"* could be labeled as {person, businessman} in S3 (investment). But he could also be labeled as {person, politician} in S1 or {person, artist, actor} in S2. Such fine-grained type representation provides more informative features for other NLP tasks. For example, since relation and event extraction pipelines rely on entity recognizer to identify possible arguments in a sentence, fine-grained argument types help distinguish hundreds or thousands of different relations and events [Ling and Weld, 2012].

Traditional named entity recognition systems adopt manually annotated corpora as training data [Nadeau and Sekine, 2007]. But the process of manually labeling a training set with large numbers of fine-grained types is too expensive and error-prone (hard for annotators to distinguish over 100 types consistently). Current fine-grained typing systems annotate training corpora automatically using KBs (i.e., *distant supervision*) [Ling and Weld, 2012, Ren et al., 2016d]. A typical workflow of distant supervision is as follows (see Fig. 5.1): (1) identify entity mentions in the documents; (2) link mentions to entities in KB; and (3) assign, to the candidate type set of each mention, all KB types of its KB-linked entity. However, existing distant supervision methods encounter the following limitations when doing automatic fine-grained typing.

• **Noisy Training Labels.** Current practice of distant supervision may introduce *label noise* to training data since it fails to take a mention's local contexts into account when assigning type

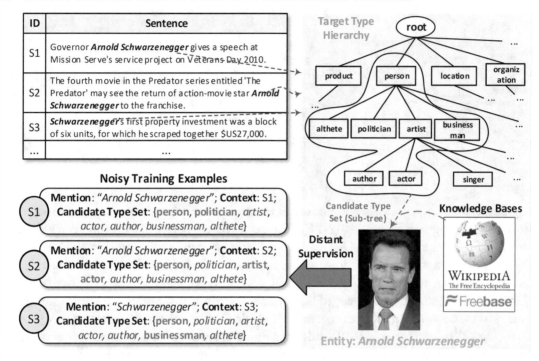

Figure 5.1: Current systems may detect *Arnold Schwarzenegger* in sentences S1–S3 and assign the same types to all (listed within braces), when only some types are correct for context (blue labels within braces).

labels (e.g., see Fig. 5.1). Many previous studies ignore the label noises which appear in a majority of training mentions (see Table 5.1, row (1)), and assume *all* types obtained by distant supervision are "correct" [Ling and Weld, 2012, Yogatama et al., 2015]. The noisy labels may mislead the trained models and cause negative effect. A few systems try to denoise the training corpora using simple pruning heuristics such as deleting mentions with conflicting types [Gillick et al., 2014]. However, such strategies significantly reduce the size of training set (Table 5.1, rows (2a–c)) and lead to performance degradation (later shown in our experiments). The larger the target type set, the more severe the loss.

• **Type Correlation.** Most existing methods [Ling and Weld, 2012, Yogatama et al., 2015] treat every type label in a training mention's candidate type set *equally* and *independently* when learning the classifiers but ignore the fact that types in the given hierarchy are semantically correlated (e.g., actor is more relevant to singer than to politician). As a consequence, the learned classifiers may bias toward popular types but perform poorly on infrequent types since training data on infrequent types is scarce. Intuitively, one should pose smaller penalty on types which are semantically more relevant to the true types. For example, in Fig. 5.1 singer should receive a smaller penalty than `politician` does, by knowing that `actor` is a true type for "*Arnold Schwarzenegger*" in S2. This provides classifiers with additional information to distinguish between two types, especially those infrequent ones.

Table 5.1: A study of label noise. (1): %mentions with multiple *sibling types* (e.g., `actor`, `singer`); (2a)–(2c): %mentions deleted by the three pruning heuristics [Gillick et al., 2014] (see Section 5.4), for three experiment datasets and NYT annotation corpus [Dunietz and Gillick, 2014].

Dataset	Wiki	OntoNotes	BBN	NYT
# of target types	113	89	47	446
(1) Noisy mentions (%)	27.99	25.94	22.32	51.81
(2a) Sibling pruning (%)	23.92	16.09	22.32	39.26
(2b) Min. pruning (%)	28.22	8.09	3.27	32.75
(2c) All pruning (%)	45.99	23.45	25.33	61.12

In this chapter, we approach the problem of *automatic fine-grained entity typing* as follows. (1) Use different objectives to model training mentions with correct type labels and mentions with noisy labels, respectively. (2) Design a novel *partial-label loss* to model true types within the noisy candidate type set which requires only the "*best*" candidate type to be relevant to the training mention, and progressively estimate the best type by leveraging various text features extracted for the mention. (3) Derive type correlation based on two signals: (i) the given type hierarchy; and (ii) the shared entities between two types in KB, and incorporate the correlation so induced by enforcing *adaptive margins* between different types for mentions in the training set. To integrate these ideas, we develop a novel embedding-based framework called **AFET**. First, it uses distant supervision to obtain candidate types for each mention, and extract a variety of text features from the mentions themselves and their local contexts. Mentions are partitioned into a "clean" set and a "noisy" set based on the given type hierarchy. Second, we embed mentions and types jointly into a low-dimensional space, where, in that space, objects (i.e., features and types) that are semantically close to each other also have similar representations. In the proposed objective, an adaptive margin-based rank loss is proposed to model the set of clean mentions to capture type correlation, and a partial-label rank loss is formulated to model the "best" candidate type for each noisy mention. Finally, with the learned embeddings (i.e., mapping matrices), one can predict the type-path for each mention in the test set in a top-down manner, using its text features. The major contributions of this chapter are as follows.

1. We propose an automatic fine-grained entity typing framework, which reduces label noise introduced by distant supervision and incorporates type correlation in a principle way.

2. A novel optimization problem is formulated to jointly embed entity mentions and types to the same space. It models noisy type set with a partial-label rank loss and type correlation with adaptive-margin rank loss.

3. We develop an iterative algorithm for solving the joint optimization problem efficiently.

4. Experiments with three public datasets demonstrate that AFET achieves significant improvement over the state of the art.

5.2 PRELIMINARIES

Our task is to automatically uncover the type information for entity mentions (i.e., token spans representing entities) in natural language sentences. The task takes a document collection $\mathcal{D}$ (automatically labeled using a KB Ψ in conjunction with a target type hierarchy $\mathcal{Y}$) as input and predicts a type-path in $\mathcal{Y}$ for each mention from the test set $\mathcal{D}_t$.

Type Hierarchy and Knowledge Base. Two key factors in distant supervision are the target type hierarchy and the KB. A *type hierarchy*, $\mathcal{Y}$, is a tree where nodes represent types of interests from Ψ. Previous studies manually create several clean type hierarchies using types from Freebase [Ling and Weld, 2012] or WordNet [Yosef et al., 2012]. In this study, we adopt the existing hierarchies constructed using Freebase types.[1] To obtain types for entities $\mathcal{E}_\Psi$ in Ψ, we use the human-curated entity-type facts in Freebase, denoted as $\mathcal{F}_\Psi = \{(e, y)\} \subset \mathcal{E}_\Psi \times \mathcal{Y}$.

Automatically Labeled Training Corpora. There exist publicly available labeled corpora such as Wikilinks [Singh et al., 2012] and ClueWeb [Gabrilovich et al., 2013]. In these corpora, entity mentions are identified and mapped to KB entities using anchor links. In specific domains (e.g., product reviews) where such public corpora are unavailable, one can utilize distant supervision to automatically label the corpus [Ling and Weld, 2012]. Specifically, an entity linker will detect mentions m_i and map them to one or more entity e_i in $\mathcal{E}_\Psi$. Types of e_i in KB are then associated with m_i to form its type set $\mathcal{Y}_i$, i.e., $\mathcal{Y}_i = \{y \mid (e_i, y) \in \mathcal{F}_\Psi, y \in \mathcal{Y}\}$. Formally, a training corpus $\mathcal{D}$ consists of a set of extracted *entity mentions* $\mathcal{M} = \{m_i\}_{i=1}^N$, the *context* (e.g., sentence, paragraph) of each mention $\{c_i\}_{i=1}^N$, and the candidate *type sets* $\{\mathcal{Y}_i\}_{i=1}^N$ for each mention. We represent $\mathcal{D}$ using a set of triples $\mathcal{D} = \{(m_i, c_i, \mathcal{Y}_i)\}_{i=1}^N$.

Problem Description. For each test mention, we aim to predict the correct type-path in $\mathcal{Y}$ based on the *mention's context*. More specifically, the test set $\mathcal{T}$ is defined as a set of mention-context pairs (m, c), where mentions in $\mathcal{T}$ (denoted as $\mathcal{M}_t$) are extracted from their sentences using existing extractors such as named entity recognizer [Finkel et al., 2005]. We denote the gold type-path for a test mention m as $\mathcal{Y}^*$. This work focuses on learning a typing model from the noisy training corpus D, and estimating $\mathcal{Y}^*$ from $\mathcal{Y}$ for each test mention m (in set $\mathcal{M}_t$), based on mention m, its context c, and the learned model.

---

[1]We use the Freebase dump as of June 30, 2015.

Framework Overview. At a high level, the AFET framework (see also Fig. 5.2) learns low-dimensional representations for entity types and text features, and infers type-paths for test mentions using the learned embeddings. It consists of the following steps.

1. Extract text features for entity mentions in training set $\mathcal{M}$ and test set $\mathcal{M}_t$ using their surface names as well as the contexts. (Section 5.3.1).

2. Partition training mentions $\mathcal{M}$ into a clean set (denoted as $\mathcal{M}_c$) and a noisy set (denoted as $\mathcal{M}_n$) based on their candidate type sets (Section 5.3.2).

3. Perform joint embedding of entity mentions $\mathcal{M}$ and type hierarchy $\mathcal{Y}$ into the same low-dimensional space where, in that space, close objects also share similar types (Sections 5.3.3–5.3.6).

4. For each test mention m, estimate its type-path $\mathcal{Y}^*$ (on the hierarchy $\mathcal{Y}$) in a top-down manner using the learned embeddings (Section 5.3.6).

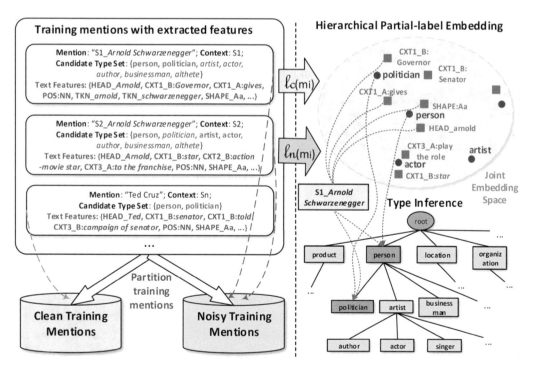

Figure 5.2: Framework Overview of AFET.

5.3   THE AFET FRAMEWORK

This section introduces the proposed framework and formulates an optimization problem for learning embeddings of text features and entity types jointly.

5.3.1   TEXT FEATURE GENERATION

We start with a representation of entity mentions. To capture the shallow syntax and distributional semantics of a mention $m_i \in \mathcal{M}$, we extract various features from both m_i itself and its context c_i. Table 5.2 lists the set of text features used in this work, which is similar to those used in Ling and Weld [2012] and Yogatama et al. [2015]. We denote the set of M unique features extracted from $\mathcal{D}$ as $\mathcal{F} = \{f_j\}_{j=1}^{M}$.

Table 5.2: Text features used in this chapter. "*Turing Machine*" is used as an example mention from "*The band's former drummer Jerry Fuchs—who was also a member of Maserati, **Turing Machine** and The Juan MacLean—died after falling down an elevator shaft.*"

Feature	Description	Example
Head	Syntactic head token of the mention	"HEAD_Turing"
Token	Tokens in the mention	"Turing", "Machine"
POS	Part-of-speech tag of tokens in the mention	"NN"
Character	All character trigrams in the head of the mention	":tu", "tur", ..., "ng:"
Word Shape	Word shape of the tokens in the mention	"Aa" for "Turing"
Length	Number of tokens in the mention	"2"
Context	Unigrams/bigrams before and after the mention	"CXT_B:Maserati ,", "CXT_A:and the"
Brown Cluster	Brown cluster ID for the head token (learned using $\mathcal{D}$)	"4_1100", "8_1101111"
Dependency	Stanford syntactic dependency [Manning et al., 2014] associated with the head token	"GOV:nn", "GOV:turing"

5.3.2   TRAINING SET PARTITION

A training mention m_i (in set $\mathcal{M}$) is considered as a "clean" mention if its candidate type set obtained by distant supervision (i.e., $\mathcal{Y}_i$) is not ambiguous, i.e., candidate types in $\mathcal{Y}_i$ can form a *single* path in tree $\mathcal{Y}$. Otherwise, a mention is considered as "noisy" mention if its candidate types form *multiple* type-paths in $\mathcal{Y}$. Following the above hypothesis, we judge each mention m_i (in set $\mathcal{M}$) and place it in either the "clean" set $\mathcal{M}_c$, or the "noisy" set $\mathcal{M}_n$. Finally, we have $\mathcal{M} = \mathcal{M}_c \cup \mathcal{M}_n$.

5.3.3 THE JOINT MENTION-TYPE MODEL

We propose to learn mappings into low-dimensional vector space, where both entity mentions and type labels (in the training set) are represented, and in that space, *two objects are embedded close to each other if and only if they share similar types*. In doing so, we later can derive the representation of a test mention based on its text features and the learned mappings. Mapping functions for entity mentions and entity type labels are different as they have different representations in the *raw* feature space, but are jointly learned by optimizing a global objective of interests to handle the aforementioned challenges.

Each entity mention $m_i \in M$ can be represented by a M-dimensional feature vector $\mathbf{m}_i \in \mathbb{R}^M$, where $m_{i,j}$ is the number of occurrences of feature f_j (in set $\mathcal{F}$) for m_i. Each type label $y_k \in \mathcal{Y}$ is represented by a K-dimensional binary indicator vector $\mathbf{y}_k \in \{0,1\}^K$, where $\mathbf{y}_{k,k} = 1$, and 0 otherwise.

Specifically, we aim to learn a mapping function from the mention's feature space to a low-dimensional vector space, i.e., $\Phi_{\mathcal{M}}(\mathbf{m}_i) : \mathbb{R}^M \mapsto \mathbb{R}^d$ and a mapping function from type label space to the same low-dimensional space, i.e., $\Phi_{\mathcal{Y}}(\mathbf{y}_k) : \mathbb{R}^K \mapsto \mathbb{R}^d$. In this work, we adopt linear maps, as similar to the mapping functions used in Weston et al. [2011]:

$$\Phi_{\mathcal{M}}(\mathbf{m}_i) = tU\mathbf{m}_i; \quad \Phi_{\mathcal{Y}}(\mathbf{y}_k) = tV\mathbf{y}_k, \tag{5.1}$$

where $tU \in \mathbb{R}^{d \times M}$ and $tV \in \mathbb{R}^{d \times K}$ are the projection matrices for mentions and type labels, respectively.

5.3.4 MODELING TYPE CORRELATION

In type hierarchy (tree) $\mathcal{Y}$, types closer to each other (i.e., shorter path) tend to be more related (e.g., `actor` is more related to `artist` than to `person` in the right column of Fig. 5.2). In KB Ψ, types assigned to similar sets of entities should be more related to each other than those assigned to quite different entities [Jiang et al., 2015] (e.g., `actor` is more related to `director` than to `author` in the left column of Fig. 5.3). Thus, type correlation between y_k and $y_{k'}$ (denoted as $w_{kk'}$) can be measured either using the *one over the length of shortest path in* $\mathcal{Y}$, or using the *normalized number of shared entities in KB*, which is defined as follows:

$$w_{kk'} = \left(|\mathcal{E}_k \cap \mathcal{E}_{k'}|/|\mathcal{E}_k| + |\mathcal{E}_k \cap \mathcal{E}_{k'}|/|\mathcal{E}_{k'}|\right)/2. \tag{5.2}$$

Although a shortest path is efficient to compute, its accuracy is limited—it is not always true that a type (e.g., `athlete`) is more related to its parent type (i.e., `person`) than to its sibling types (e.g., `coach`), or that all sibling types are equally related to each other (e.g., `actor` is more related to `director` than to `author`). We later compare these two methods in our experiments.

With the type correlation computed, we propose to apply *adaptive* penalties on different negative type labels (for a training mention), instead of treating all of the labels *equally* as in most existing work [Weston et al., 2011]. The hypothesis is intuitive: given the positive type labels for a mention, we force the negative type labels which are related to the positive type labels to receive smaller penalty. For example, in the right column of Fig. 5.3, negative label `businessman` receives a smaller penalty (i.e., margin) than `athele` does, since `businessman` is more related to `politician`.

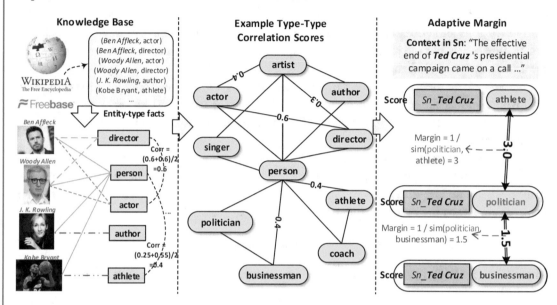

Figure 5.3: An illustration of KB-based type correlation computation, and the proposed adaptive margin.

> ### Hypothesis 5.1: Adaptive Margin
>
> For a mention, if a negative type is correlated to a positive type, the margin between them should be smaller.

We propose an adaptive-margin rank loss to model the set of "clean" mentions (i.e., $\mathcal{M}_c$), based on the above hypothesis. The intuition is simple: for each mention, rank all the positive types ahead of negative types, where the ranking score is measured by similarity between mention and type. We denote $f_k(m_i)$ as the similarity between (m_i, y_k) and is defined as the inner product of $\Phi_{\mathcal{M}}(\mathbf{m}_i)$ and $\Phi_{\mathcal{Y}}(\mathbf{y}_k)$:

$$\ell_c(m_i, \mathcal{Y}_i, \overline{\mathcal{Y}}_i) = \sum_{y_k \in \mathcal{Y}_i} \sum_{y_{\bar{k}} \in \overline{\mathcal{Y}}_i} L\left\lfloor \mathrm{rank}_{y_k}\left(f(m_i)\right) \right\rfloor \Theta_{i,k,\bar{k}};$$

$$\Theta_{i,k,\bar{k}} = \max\left\{0, \gamma_{k,\bar{k}} - f_k(m_i) + f_{\bar{k}}(m_i)\right\};$$

$$\mathrm{rank}_{y_k}\left(f(m_i)\right) = \sum_{y_{\bar{k}} \in \overline{\mathcal{Y}}_i} \mathbf{1}\left(\gamma_{k,\bar{k}} + f_{\bar{k}}(m_i) > f_k(m_i)\right).$$

Here, $\gamma_{k,\bar{k}}$ is the adaptive margin between positive type k and negative type $\bar{k}$, which is defined as $\gamma_{k,\bar{k}} = 1 + 1/(w_{k,\bar{k}} + \alpha)$ with a smooth parameter α. $L(x) = \sum_{i=1}^{x} \frac{1}{i}$ transforms rank to a weight, which is then multiplied to the max-margin loss $\Theta_{i,k,\bar{k}}$ to optimize precision at x [Weston et al., 2011].

5.3.5   MODELING NOISY TYPE LABELS

True type labels for noisy entity mentions $\mathcal{M}_n$ (i.e., mentions with ambiguous candidate types in the given type hierarchy) in each sentence are not available in KBs. To effectively model the set of noisy mentions, we propose not to treat *all* candidate types (i.e., $\{\mathcal{Y}_i\}$ as true labels. Instead, we model the "true" label among the candidate set as *latent value*, and try to infer that using text features (Fig. 5.4).

> **Hypothesis 5.2: Partial-Label Loss**
>
> For a noisy mention, the maximum score associated with its candidate types should be greater than the scores associated with any other non-candidate types.

We extend the partial-label loss in Nguyen and Caruana [2008] (used to learn linear classifiers) to enforce Hypothesis 5.2, and integrate with the adaptive margin to define the loss for m_i (in set $\mathcal{M}_n$):

$$\ell_n(m_i, \mathcal{Y}_i, \overline{\mathcal{Y}}_i) = \sum_{\bar{k} \in \overline{\mathcal{Y}}_i} L\left\lfloor \mathrm{rank}_{y_{k*}}\left(f(m_i)\right) \right\rfloor \Omega_{i,\bar{k}};$$

$$\Omega_{i,k} = \max\left\{0, \gamma_{k*,\bar{k}} - f_{k*}(m_i) + f_{\bar{k}}(m_i)\right\};$$

$$\mathrm{rank}_{y_{k*}}\left(f(m_i)\right) = \sum_{y_{\bar{k}} \in \overline{\mathcal{Y}}_i} \mathbf{1}\left(\gamma_{k*,\bar{k}} + f_{\bar{k}}(m_i) > f_{k*}(m_i)\right),$$

where we define . $y_{k*} = \mathrm{argmax}_{y_k \in \mathcal{Y}_i} f_k(m_i)$ and $y_{\bar{k}*} = \mathrm{argmax}_{y_k \in \overline{\mathcal{Y}}_i} f_k(m_i)$.

Minimizing ℓ_n encourages a *large margin* between the maximum scores $\max_{y_k \in \mathcal{Y}_i} f_{y_k}(m_i)$ and $\max_{y_k \in \overline{\mathcal{Y}}_i} f_{y_k}(m_i)$. This forces m_i to be embedded closer to the most "relevant" type in the noisy candidate type set, i.e., $y^* = \mathrm{argmax}_{y_k \in \mathcal{Y}_i} f_{y_k}(m_i)$, than to *any other* non-candidate types

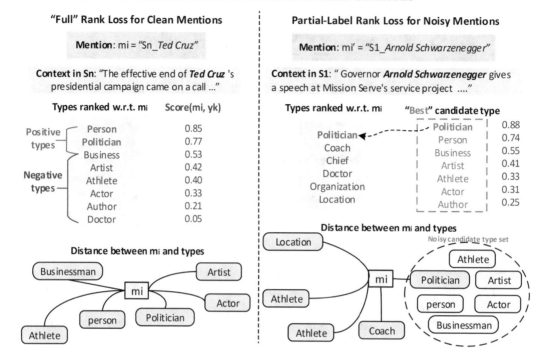

Figure 5.4: An illustration of the partial-label rank loss.

(i.e., Hypothesis 5.2). This contrasts sharply with multi-label learning [Yosef et al., 2012], where a large margin is enforced between *all* candidate types and non-candidate types without considering noisy types.

5.3.6 HIERARCHICAL PARTIAL-LABEL EMBEDDING

Our goal is to embed the heterogeneous graph G into a d-dimensional vector space, following the three proposed hypotheses in the section. Intuitively, one can *collectively* minimize the objectives of the two kinds of loss functions ℓ_c and ℓ_n, across all the training mentions. To achieve the goal, we formulate a joint optimization problem as follows:

$$\min_{tU,\, tV} \mathcal{O} = \sum_{m_i \in \mathcal{M}_c} \ell_c(m_i, \mathcal{Y}_i, \overline{\mathcal{Y}}_i) + \sum_{m_i \in \mathcal{M}_n} \ell_n(m_i, \mathcal{Y}_i, \overline{\mathcal{Y}}_i).$$

We use an alternative minimization algorithm based on block-wise coordinate descent [Tseng, 2001] to *jointly* optimize the objective $\mathcal{O}$. One can also apply stochastic gradient descent to do online update.

Type Inference. With the learned mention embeddings $\{\mathbf{u}_i\}$ and type embeddings $\{\mathbf{v}_k\}$, we perform top-down search in the given type hierarchy $\mathcal{Y}$ to estimate the correct type-path y_i^*.

Starting from the tree's root, we recursively find the best type among the children types by measuring the dot product of the corresponding mention and type embeddings, i.e., $\text{sim}(\mathbf{u}_i, \mathbf{v}_k)$. The search process stops when we reach a leaf type, or the similarity score is below a pre-defined threshold $\eta > 0$.

5.4   EXPERIMENTS

5.4.1   DATA PREPARATION

Datasets. Our experiments use three public datasets. (1) **Wiki** [Ling and Weld, 2012]: consists of 1.5M sentences sampled from Wikipedia articles; (2) **OntoNotes** [Weischedel et al., 2011]: consists of 13,109 news documents where 77 test documents are manually annotated [Gillick et al., 2014]; and (3) **BBN** [Weischedel and Brunstein, 2005]: consists of 2,311 *Wall Street Journal* articles which are manually annotated using 93 types. Statistics of the datasets are shown in Table 5.3.

Training Data. We followed the process in Ling and Weld [2012] to generate training data for the Wiki dataset. For the BBN and OntoNotes datasets, we used DBpedia Spotlight[2] for entity linking. We discarded types which cannot be mapped to Freebase types in the BBN dataset (47 of 93).

Table 5.3: Statistics of the datasets

Dataset	Wiki	OntoNotes	BBN
# Types	113	89	47
# Documents	780,549	13,109	2,311
# Sentences	1.51M	143,709	48,899
# Training mentions	2.69M	223,342	109,090
# Ground-truth mentions	563	9,604	121,001
# Features	644,860	215,642	125,637
# Edges in graph	87M	5.9M	2.9M

Table 5.2 lists the set of features used in our experiments, which are similar to those used in Ling and Weld [2012] and Yogatama et al. [2015] except for topics and ReVerb patterns. We used a six-word window to extract context unigrams and bigrams for each mention (three words on the left and the right). We applied the Stanford CoreNLP tool [Manning et al., 2014] to get POS tags and dependency structures. The word clusters were derived for each corpus using the Brown clustering algorithm.[3] Features for a mention is represented as a binary indicator vector

[2]http://spotlight.dbpedia.org/
[3]https://github.com/percyliang/brown-cluster

where the dimensionality is the number of features derived from the corpus. We discarded the features which occur only once in the corpus. The number of features generated for each dataset is shown in Table 5.3.

5.4.2   EVALUATION SETTINGS

For the Wiki and OntoNotes datasets, we used the provided test set. Since BBN corpus is fully annotated, we followed a 80/20 ratio to partition it into training/test sets. We report Accuracy (Strict-F1), Micro-averaged F1 (Mi-F1) and Macro-averaged F1 (Ma-F1) scores commonly used in the fine-grained type problem [Ling and Weld, 2012, Yogatama et al., 2015]. Since we use the gold mention set for testing, the Accuracy (**Acc**) we reported is the same as the Strict F1.

Baselines. We compared the proposed method (AFET) and its variant with state-of-the-art typing methods, embedding methods, and partial-label learning methods: (1) **FIGER** [Ling and Weld, 2012]; (2) **HYENA** [Yosef et al., 2012]; (3) **FIGER/HYENA-Min** [Gillick et al., 2014]: removes types appearing only once in the document; (4) **ClusType** [Ren et al., 2015]: predicts types based on co-occurring relation phrases; (5) **HNM** [Dong et al., 2015]: proposes a hybrid neural model without hand-crafted features; (6) **DeepWalk** [Perozzi et al., 2014]: applies Deep Walk to a feature-mention-type graph by treating all nodes as the same type; (7) **LINE** [Tang et al., 2015]: uses a second-order LINE model on feature-type bipartite graph; (8) **PTE** [Tang et al., 2015]: applies the PTE joint training algorithm on feature-mention and type-mention bipartite graphs; (9) **WSABIE** [Yogatama et al., 2015]: adopts WARP loss to learn embeddings of features and types; (10) **PL-SVM** [Nguyen and Caruana, 2008]: uses a margin-based loss to handle label noise; and (11) **CLPL** [Cour et al., 2011]: uses a linear model to encourage large average scores for candidate types.

We compare AFET and its variant: (1) **AFET**: complete model with KB-induced type correlation; (2) **AFET-CoH**: with hierarchy-induced correlation (i.e., shortest path distance); (3) **AFET-NoCo**: without type correlation (i.e., all margin are "1") in the objective $\mathcal{O}$; and (4) **AFET-NoPa**: without label partial loss in the objective $\mathcal{O}$.

5.4.3   PERFORMANCE COMPARISON AND ANALYSES

Table 5.4 shows the results of AFET and its variants.

Comparison with the other typing methods. AFET outperforms both FIGER and HYENA systems, demonstrating the predictive power of the learned embeddings, and the effectiveness of modeling type correlation information and noisy candidate types. We also observe that pruning methods do not always improve the performance, since they aggressively filter out rare types in the corpus, which may lead to low Recall. ClusType is not as good as FIGER and HYENA because it is intended for coarse types and only utilizes relation phrases.

Comparison with the other embedding methods. AFET performs better than all other embedding methods. HNM does not use any linguistic features. None of the other embedding methods consider the label noise issue and treat the candidate type sets as clean. Although AFET adopts the WARP loss in WSABIE, it uses an adaptive margin in the objective to capture the type correlation information.

Comparison with partial-label learning methods. Compared with PL-SVM and CLPL, AFET obtains superior performance. PL-SVM assumes that only *one* candidate type is correct and does not consider type correlation. CLPL simply averages the model output for all candidate types, and thus may generate results biased to frequent false types. Superior performance of AFET mainly comes from modeling type correlation derived from KB.

Comparison with its variants. AFET always outperforms its variant on all three datasets. It gains performance from capturing type correlation, as well as handling type noise in the embedding process.

5.5 DISCUSSION AND CASE ANALYSIS

Example output on news articles. Table 5.5 shows the types predicted by AFET, FIGER, PTE, and WSABIE on two news sentences from OntoNotes dataset: AFET predicts fine-grained types with better accuracy (e.g., `person_title`) and avoids overly-specific predictions (e.g., `news_company`).

Testing the effect of training set size and dimension. Experimenting with the same settings for model learning, Fig. 5.5a shows the performance trend on the Wiki dataset when varying the sampling ratio (subset of mentions randomly sampled from the training set $\mathcal{D}$). Figure 5.5b analyzes the performance sensitivity of AFET with respect to d—the embedding dimension on the BBN dataset. Accuracy of AFET improves as d becomes large but the gain decreases when d is large enough.

Testing sensitivity of the tuning parameter. Figure 5.6b analyzes the sensitivity of AFET with respect to α on the BBN dataset. Performance increases as α becomes large. When α is large than 0.5, the performance becomes stable.

Testing at different type levels. Figure 5.6a reports the Ma-F1 of AFET, FIGER, PTE, and WSABIE at different levels of the target type hierarchy (e.g., person and location on level-1, politician and artist on level-2, author and actor on level-3). The results show that it is more difficult to distinguish among more fine-grained types. AFET always outperforms the other two method, and achieves a 22.36% improvement in Ma-F1, compared to FIGER on level-3 types. The gain mainly comes from explicitly modeling the noisy candidate types.

Table 5.4: Study of typing performance on the three datasets

Typing	Wiki			OntoNotes			BBN		
Method	Acc	MaF1	MiF1	Acc	MaF1	MiF1	Acc	MaF1	MiF1
CLPL [Cour et al., 2011]	0.162	0.431	0.411	0.201	0.347	0.358	0.438	0.603	0.536
PL-SVM [Nguyen and Caruana, 2008]	0.428	0.613	0.571	0.225	0.455	0.437	0.465	0.648	0.582
FIGER [Ling and Weld, 2012]	0.474	0.692	0.655	0.369	0.578	0.516	0.467	0.672	0.612
FIGER-Min [Gillick et al., 2014]	0.453	0.691	0.631	0.373	0.570	0.509	0.444	0.671	0.613
HYENA [Yosef et al., 2012]	0.288	0.528	0.506	0.249	0.497	0.446	0.523	0.576	0.587
HYENA-Min	0.325	0.566	0.536	0.295	0.523	0.470	0.524	0.582	0.595
ClusType [Ren et al., 2015]	0.274	0.429	0.448	0.305	0.468	0.404	0.441	0.498	0.573
HNM [Dong et al., 2015]	0.237	0.409	0.417	0.122	0.288	0.272	0.551	0.591	0.606
DeepWalk [Perozzi et al., 2014]	0.414	0.563	0.511	0.479	0.669	0.611	0.586	0.638	0.628
LINE [Tang et al., 2015]	0.181	0.480	0.499	0.436	0.634	0.578	0.576	0.687	0.690
PTE [Tang et al., 2015]	0.405	0.575	0.526	0.436	0.630	0.572	0.604	0.684	0.695
WSABIE [Yogatama, et al., 2015]	0.480	0.679	0.657	0.404	0.580	0.527	0.619	0.670	0.680
AFET-NoCo	0.526	0.693	0.654	0.486	0.652	0.594	0.655	0.711	0.716
AFET-NoPa	0.513	0.675	0.642	0.463	0.637	0.591	0.669	0.715	0.724
AFET-CoH	0.433	0.583	0.551	0.521	0.680	0.609	0.657	0.703	0.712
AFET	**0.533**	**0.693**	**0.664**	**0.551**	**0.711**	**0.647**	**0.670**	**0.727**	**0.735**

5.6   SUMMARY

In this chapter, we study *automatic fine-grained entity typing* and propose a *hierarchical partial-label embedding* method, AFET, that models "clean" and "noisy" mentions separately and incorporates a given type hierarchy to induce loss functions. APEFT builds on a joint optimization framework, learns embeddings for mentions and type-paths, and iteratively refines the model. Experiments on three public datasets show that AFET is effective, robust, and outperforms other comparing methods.

Table 5.5: Example output of AFET and the compared methods on two news sentences from **OntoNotes** dataset

Text	"... going to be an imminent easing of monetary policy," said Robert Dederick , chief economist at ***Northern Trust Co. in Chicago***.	...It's terrific for advertisers to know the reader will be paying more ," said Michael Drexler, ***national media director*** at Bozell Inc. ad agency.
Ground Truth	organization, company	person, person_title
FIGER	organization	organization
WSABIE	organization, company, broadcast	organization, company, news_company
PTE	organization	person
AFET	organization, company	person, person_title

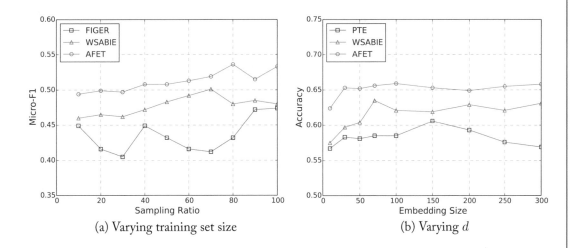

(a) Varying training set size

(b) Varying d

Figure 5.5: Performance change with respect to (a) sampling ratio of training mentions on the Wiki dataset; and (b) embedding dimension d on the BBN dataset.

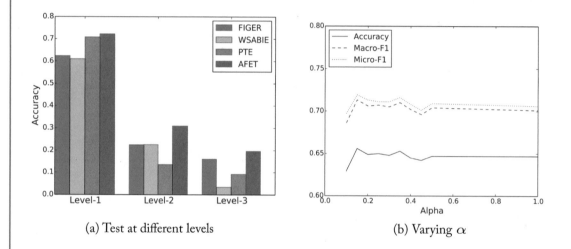

(a) Test at different levels   (b) Varying α

Figure 5.6: Performance change (a) at different levels of the type hierarchy on the OntoNotes dataset; and (b) with respect to smooth parameter α on the BBN dataset.

C H A P T E R   6

Synonym Discovery from Large Corpus

Meng Qu, *Department of Computer Science, University of Illinois at Urbana-Champaign*

People often have a variety of ways to refer to the same real-world entity, forming different *synonyms* for the entity (e.g., entity *United States* can be referred using "*America*" and "*U.S.*"). Automatic synonym discovery is an important task in text analysis and understanding, as the extracted synonyms (i.e., the alternative ways to refer to the same entity) can benefit many downstream applications [Angheluta, 2002, Wu and Weld, 2010, Xie et al., 2015, Zeng et al., 2012]. For example, by forcing synonyms of an entity to be assigned in the same topic category, one can constrain the topic modeling process and yield topic representations with higher quality [Xie et al., 2015]. Another example is in document retrieval [Voorhees, 1994], where we can leverage entity synonyms to enhance the process of query expansion, and thus improve retrieval performance.

6.1   OVERVIEW AND MOTIVATION

One straightforward approach for obtaining entity synonyms is to leverage publicly available KBs such as Freebase and WordNet, in which popular synonyms for the entities are manually curated by human crowds. However, the coverage of KBs can be rather limited, especially on some newly emerging entities, as the manual curation process entails high costs and is not scalable. For example, the entities in Freebase have only 1.1 synonyms on average. To increase the synonym coverage, we expect to automatically extract more synonyms that are not in KBs from massive, domain-specific text corpora. Many approaches address this problem through supervised [Roller et al., 2014, Wang et al., 2015, Weeds et al., 2014] or weakly supervised learning [Nakashole et al., 2012, Snow et al., 2004], which treat some manually labeled synonyms as seeds to train a synonym classifier or detect some local patterns for synonym discovery. Though quite effective in practice, such approaches still rely on careful seed selections by humans.

To retrieve training seeds automatically, recently there is a growing interest in the distant supervision strategy, which aims to automatically collect training seeds from existing KBs. The typical workflow is: (i) detect entity mentions from the given corpus, (ii) map the detected entity mentions to the entities in a given KB, and (iii) collect training seeds from the KB. Such techniques have been proved effective in a variety of applications, such as relation extraction [Mintz et al., 2009], entity typing [Ren et al., 2015], and emotion classification [Purver and Battersby, 2012]. Inspired by such strategy, a promising direction for automatic synonym discovery could be collecting training seeds (i.e., a set of synonymous strings) from KBs.

6.1.1 CHALLENGES

Although distant supervision helps collect training seeds automatically, it also poses a challenge due to the string ambiguity problem, that is, the same entity surface strings can be mapped to different entities in KBs. For example, consider the string "*Washington*" in Fig. 6.1. The "*Washington*" in the first sentence represents a state of the United States, while in the second sentence it refers to a person. As some strings like "*Washington*" have ambiguous meanings, directly inferring synonyms for such strings may lead to a set of synonyms for multiple entities. For example, the synonyms of entity *Washington* returned by current systems may contain both the state names and person names, which is not desirable. To address the challenge, instead of using ambiguous strings as queries, a better way is using some specific concepts as queries to disambiguate, such as entities in KBs.

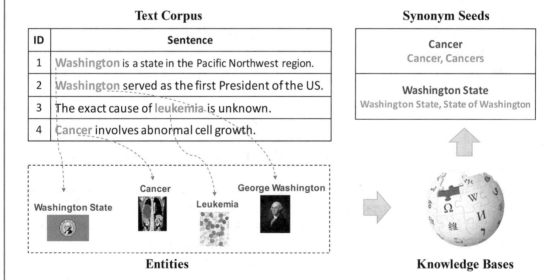

Figure 6.1: Distant supervision for synonym discovery. We link entity mentions in text corpus to knowledge base entities, and collect training seeds from KBs.

This motivates us to define a new task: *automatic synonym discovery for entities with KBs*. Given a domain-specific corpus, we aim to collect existing name strings of entities from KBs as seeds. For each query entity, the existing name strings of that entity can disambiguate the meaning for each other, and we will let them vote to decide whether a given candidate string is a synonym of the query entity. Based on that, the key task for this problem is to predict whether a pair of strings are synonymous. For this task, the collected seeds can serve as supervision to help determine the important features. However, as the synonym seeds from KBs are usually quite limited, how to use them effectively becomes a major challenge. There are broadly two kinds of efforts toward exploiting a limited number of seed examples.

The distributional-based approaches [Mikolov et al., 2013, Pennington et al., 2014, Roller et al., 2014, Wang et al., 2015, Weeds et al., 2014] consider the corpus-level statistics, and they assume strings which often appear in similar contexts are likely to be synonyms. For example, the strings "*U.S.*" and "*United States*" are usually mentioned in similar contexts, and they are the synonyms of the country *U.S.*. Based on the assumption, the distributional-based approaches usually represent strings with their distributional features, and treat the synonym seeds as labels to train a classifier, which predicts whether a given pair of strings are synonymous or not. Since most synonymous strings will appear in similar contexts, such approaches usually have high recall. However, such strategy also brings some noise, since some non-synonymous strings may also share similar contexts, such as "*U.S.*" and "*Canada*," which could be labeled as synonyms incorrectly (Fig. 6.2).

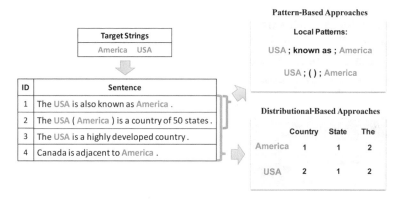

Figure 6.2: Comparison of the distributional-based and pattern-based approaches. To predict the relation of two target strings, the distributional-based approaches will analyze their distributional features, while the pattern-based approaches will analyze the local patterns extracted from sentences mentioning both strings.

Alternatively, the pattern-based approaches [Hearst, 1992, Qian et al., 2009, Snow et al., 2004, Sun and Grishman, 2010] consider the local contexts, and they infer the relation of two strings by analyzing sentences mentioning both of them. For example, from the sentence "*The United States of America is commonly referred to as America.*", we can infer that "*United States of America*" and "*America*" have the synonym relation, while the sentence "*The U.S. is adjacent to Canada*" may imply that "*U.S.*" and "*Canada*" are not synonymous. To leverage this observation, the pattern-based approaches will extract some textual patterns from sentences in which two synonymous strings co-occur, and discover more synonyms with the learned patterns. Different from the distributional-based approaches, the pattern-based approaches can treat the patterns as concrete evidences to support the discovered synonyms, which are more convincing and interpretable. However, as many synonymous strings will not be co-mentioned in any sentences, such approaches usually suffer from low recall.

6.1.2   PROPOSED SOLUTION

Ideally, we would wish to combine the merits of both approaches, and in this chapter we propose such a solution named DPE (distributional- and pattern-integrated embedding). Our framework consists of a distributional module and a pattern module. The distributional module predicts the synonym relation from the *global* distributional features of strings; while in the pattern module, we aim to discover synonyms from the *local* contexts. Both modules are built on top of some string embeddings, which preserve the semantic meanings of strings. During training, both modules will treat the embeddings as features for synonym prediction, and in turn update the embeddings based on the supervision from synonym seeds. The string embeddings are shared across the modules, and therefore each module can leverage the knowledge discovered by the other module to improve the learning process.

To discover missing synonyms for an entity, one may directly rank all candidate strings with both modules. However, such strategy can have high time costs, as the pattern module needs to extract and analyze all sentences mentioning a pair of given strings when predicting their relation. To speed up synonym discoveries, our framework will first utilize the distributional module to rank all candidate strings, and extract a set of top ranked candidates as high-potential ones. After that, we will re-rank the high-potential candidates with both modules, and treat the top-ranked candidates as the discovered synonyms.

6.2   THE DPE FRAMEWORK

In this section, we introduce our approach DPE for entity synonym discovery with KBs. To infer the synonyms of a query entity, we leverage its name strings collected from KBs to disambiguate the meaning for each other, and let them vote to decide whether a given candidate string is a synonym of the query entity. Therefore, the key task for this problem is to predict whether a pair of strings are synonymous or not. For this task, the synonym seeds collected from KBs can serve as supervision to guide the learning process. However, as the number of synonym seeds is usually small, how to leverage them effectively is quite challenging. Existing approaches either train a synonym classifier with the distributional features, or learn some textual patterns for synonym discovery, which cannot exploit the seeds sufficiently.

To address this challenge, our framework naturally integrates the distributional-based approaches and the pattern-based approaches. Specifically, our framework consists of a distributional module and a pattern module. Given a pair of target strings, the distributional module predicts the synonym relation from the global distributional features of each string, while the pattern module considers the local contexts mentioning both target strings. During training, both modules will mutually enhance each other. At the inference stage, we will leverage both modules to find high-quality synonyms for the query entities.

Framework Overview. The overall framework of DPE (Fig. 6.3) is summarized below.

1. Detect entity mentions in the given text corpus and link them to entities in the given KB. Collect synonym seeds from KBs as supervision.

2. Jointly optimize the distributional and the pattern modules. The distributional module predicts synonym relations with the global distributional features, while the pattern module considers the local contexts mentioning both target strings.

3. Discover missing synonyms for the query entities with both the distributional module and the pattern module.

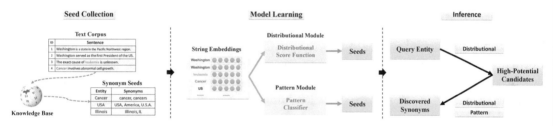

Figure 6.3: Framework overview.

6.2.1 SYNONYM SEED COLLECTION

To automatically collect synonym seeds, our approach will first detect entity mentions (strings that represent entities) in the given text corpus and link them to entities in the given KB. After that, we will retrieve the existing synonyms in KBs as our training seeds. An illustrative example is presented in Fig. 6.1.

6.2.2 JOINT OPTIMIZATION PROBLEM

After extracting synonym seeds from KBs, we formulate an optimization framework to jointly learn the distributional module and pattern module.

To preserve the semantic meanings of different strings, our framework introduces a low-dimensional vector (embedding) to represent each entity surface string (i.e., strings that are linked to entities in KBs) and each unlinkable string (i.e., words and phrases that are not linked to any entities). For the same strings that linked to different entities, as they have different semantic meanings, we introduce different embeddings for them. For example, the string "*Washington*" can be linked to a state or a person, and we use two embeddings to represent *Washington* (state) and *Washington* (person) respectively.

The two modules of our framework are built on top of these string embeddings. Specifically, both modules treat the embeddings as features for synonym prediction, and in turn update the embeddings based on the supervision from the synonym seeds, which may bring stronger predictive abilities to the learned embeddings. Meanwhile, since the string embeddings are shared between the two modules, each module is able to leverage the knowledge discovered by the other module, so that the two modules can mutually enhance to improve the learning process.

The overall objective of our framework is summarized as follows:

$$O = O_D + O_P, \tag{6.1}$$

where O_D is the objective of the distributional module and O_P is the objective of the pattern module. Next, we introduce the details of each module.

6.2.3   DISTRIBUTIONAL MODULE

The distributional module of our framework considers the global distributional features for synonym discovery. The module consists of an unsupervised part and a supervised part. In the unsupervised part, a co-occurrence network encoding the distributional information of strings will be constructed, and we try to preserve the distributional information into the string embeddings. Meanwhile in the supervised part, the synonym seeds will be used to learn a distributional score function, which takes string embeddings as features to predict whether two strings are synonymous or not.

Unsupervised Part. In the unsupervised part, we first construct a co-occurrence network between different strings, which captures their distributional information. Formally, all strings (i.e., entity surface strings and other unlinkable strings) within a sliding window of a certain size w in the text corpus are considered to be co-occurring with each other. The weight for each pair of strings in the co-occurrence network is defined as their co-occurrence count.

Supervised Part. The unsupervised part of the distributional module can effectively preserve the distributional information of strings into the learned string embeddings. In the supervised part, we will utilize the collected synonym seeds to train a distributional score function, which treats the string embeddings as features to predict whether two strings have the synonym relation or not.

6.2.4   PATTERN MODULE

For a pair of target strings, the pattern module of our framework predicts their relation from the sentences mentioning both of them (Fig. 6.4). We achieve this by extracting a pattern from each of such sentences, and collecting some lexical features and syntactic features to represent each pattern. Based on the extracted features, a pattern classifier is trained to predict whether a pattern expresses the synonym relation between the target strings. Finally, we will integrate all prediction results from these patterns to decide the relation of the target strings.

Sentence	Illinois , which is also called IL , is a state in the US .
Pattern	(ENT NNP nsubj) (called VBN acl:relcl) (ENT NN xcomp)
Lexical Features	Embedding[called]
Syntactic Features	NNP VBN NNP (NNP,VBN) (VBN,NNP) nsubj acl:relcl xcomp (nsubj,acl:relcl) (acl,xcomp)

Sentence	Michigan , also known as MI , consists of two peninsulas .
Pattern	(ENT NNP nsubj) (known VBN acl) (ENT NNP xcomp)
Lexical Features	Embedding[known]
Syntactic Features	NNP VBN NNP (NNP,VBN) (VBN,NNP) nsubj acl xcomp (nsubj,acl) (acl,xcomp)

Figure 6.4: Examples of patterns and their features. For a pair of target strings (red ones) in each sentence, we define the pattern as the <token, POS tag, dependency label> triples in the shortest dependency path. We collect both lexical features and syntactic features for pattern classification.

6.3   EXPERIMENT

1. Comparing DPE with other baseline approaches. Table 6.1, Table 6.2, and Figure 6.5 present the results on the warm-start and cold-start settings. In both settings, we see that the pattern-based approach Patty does not perform well, and our proposed approach DPE significantly outperforms Patty. This is because most synonymous strings will never co-appear in any sentences, leading to the low recall of Patty. Also, many patterns discovered by Patty are not so reliable, which may harm the precision of the discovered synonyms. DPE addresses this problem by incorporating the distributional information, which can effectively complement and regulate the pattern information, leading to higher recall and precision.

Comparing DPE with the distributional based approaches (word2vec, GloVe, PTE, RKPM), DPE still significantly outperforms them. The performance gains mainly come from: (1) we exploit the co-occurrence observation during training, which enables us to better capture the semantic meanings of different strings; and (2) we incorporate the pattern information to improve the performances.

2. Comparing DPE with its variants. To better understand why DPE achieves better results, we also compare DPE with several variants. From Table 6.1 and Table 6.2, we see that in most cases, the distributional module of our approach (DPE-NoP) can already outperform the best

Table 6.1: Quantitative results on the warm-start setting

Algorithm	Wiki + Freebase						PubMed + UMLS						NYT + Freebase					
	P@1	R@1	F1@1	P@5	R@5	F1@5	P@1	R@1	F1@1	P@5	R@5	F1@5	P@1	R@1	F1@1	P@5	R@5	F1@5
Patty	0.102	0.075	0.086	0.049	0.167	0.076	0.352	0.107	0.164	0.164	0.248	0.197	0.101	0.081	0.090	0.038	0.141	0.060
SVM	0.508	0.374	0.431	0.273	0.638	0.382	0.696	0.211	0.324	0.349	0.515	0.416	0.481	0.384	0.427	0.248	0.616	0.354
word2vec	0.387	0.284	0.328	0.247	0.621	0.353	0.784	0.238	0.365	0.464	0.659	0.545	0.367	0.293	0.326	0.216	0.596	0.317
GloVe	0.254	0.187	0.215	0.104	0.316	0.156	0.536	0.163	0.250	0.279	0.417	0.334	0.203	0.162	0.180	0.084	0.283	0.130
PTE	0.445	0.328	0.378	0.252	0.612	0.357	0.800	0.243	0.373	0.476	0.674	0.558	0.456	0.364	0.405	0.233	0.606	0.337
RKPM	0.500	0.368	0.424	0.302	0.681	0.418	0.804	0.244	0.374	0.480	0.677	0.562	0.506	0.404	0.449	0.302	0.707	0.423
DPE-NoP	0.641	0.471	0.543	0.414	0.790	0.543	0.816	0.247	0.379	0.532	0.735	0.617	0.532	0.424	0.472	0.305	0.687	0.422
DPE-TwoStep	0.684	0.503	0.580	0.417	0.782	0.544	0.836	0.254	0.390	0.538	0.744	0.624	0.557	0.444	0.494	0.344	0.768	0.475
DPE	**0.727**	**0.534**	**0.616**	**0.465**	**0.816**	**0.592**	**0.872**	**0.265**	**0.406**	**0.549**	**0.755**	**0.636**	**0.570**	**0.455**	**0.506**	**0.366**	**0.788**	**0.500**

Table 6.2: Quantitative results on the cold-start setting

Algorithm	Wiki + Freebase						PubMed + UMLS						NYT + Freebase					
	P@1	R@1	F1@1	P@5	R@5	F1@5	P@1	R@1	F1@1	P@5	R@5	F1@5	P@1	R@1	F1@1	P@5	R@5	F1@5
Patty	0.131	0.056	0.078	0.065	0.136	0.088	0.413	0.064	0.111	0.191	0.148	0.167	0.125	0.054	0.075	0.062	0.132	0.084
SVM	0.371	0.158	0.222	0.150	0.311	0.202	0.707	0.110	0.193	0.381	0.297	0.334	0.347	0.150	0.209	0.165	0.347	0.224
word2vec	0.411	0.175	0.245	0.196	0.401	0.263	0.627	0.098	0.170	0.408	0.318	0.357	0.361	0.156	0.218	0.151	0.317	0.205
GloVe	0.251	0.107	0.150	0.105	0.221	0.142	0.480	0.075	0.130	0.264	0.206	0.231	0.181	0.078	0.109	0.084	0.180	0.115
PTE	0.474	0.202	0.283	0.227	0.457	0.303	0.647	0.101	0.175	0.389	0.303	0.341	0.403	0.174	0.243	0.166	0.347	0.225
RKPM	0.480	0.204	0.286	0.227	0.455	0.303	0.700	0.109	0.189	0.447	0.348	0.391	0.403	0.186	0.255	0.170	0.353	0.229
DPE-NoP	0.491	0.209	0.293	0.246	0.491	0.328	0.700	0.109	0.189	0.456	0.355	0.399	0.417	0.180	0.251	0.180	0.371	0.242
DPE-TwoStep	0.537	0.229	0.321	0.269	0.528	0.356	0.720	0.112	0.194	0.477	0.372	0.418	0.431	0.186	0.260	0.183	0.376	0.246
DPE	**0.646**	**0.275**	**0.386**	**0.302**	**0.574**	**0.396**	**0.753**	**0.117**	**0.203**	**0.500**	**0.389**	**0.438**	**0.486**	**0.201**	**0.284**	**0.207**	**0.400**	**0.273**

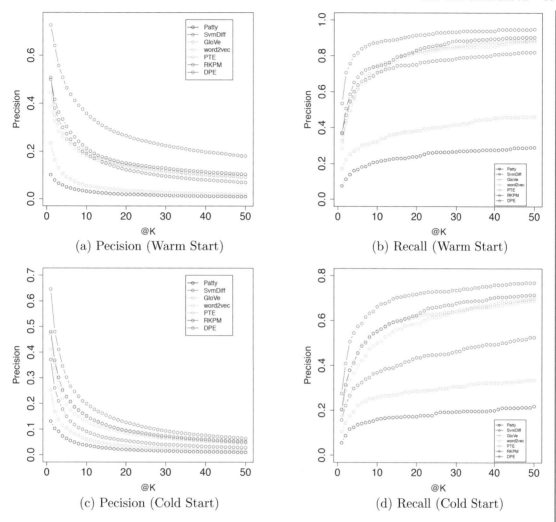

(a) Pecision (Warm Start)

(b) Recall (Warm Start)

(c) Pecision (Cold Start)

(d) Recall (Cold Start)

Figure 6.5: Precision and recall at different positions on the Wiki dataset.

baseline approach RKPM. This is because we utilize the co-occurrence observation in our distributional module, which helps us capture the semantic meanings of strings more effectively. By separately training the pattern module after the distributional module, and using both modules for synonym discovery (DPE-TwoStep), we see that the results are further improved, which demonstrates that the two modules can indeed mutually complement each other for synonym discovery. If we jointly train both modules (DPE), we obtain even better results, which shows that our proposed joint optimization framework can benefit the training process and therefore helps achieve better results.

6.4   SUMMARY

Recognizing entity synonyms from text has become a crucial task in many entity-leveraging applications. However, discovering entity synonyms from domain-specific text corpora (e.g., news articles, scientific papers) is rather challenging. Current systems take an entity name string as input to find out other names that are synonymous, ignoring the fact that often times a name string can refer to multiple entities (e.g., "apple" could refer to both *Apple Inc* and the fruit *apple*). Moreover, most existing methods require training data manually created by domain experts to construct supervised-learning systems. In this chapter, we study the problem of automatic synonym discovery with KBs, that is, identifying synonyms for *knowledge base entities* in a given domain-specific corpus. The manually curated synonyms for each entity stored in a KB not only form a set of name strings to *disambiguate* the meaning for each other, but also can serve as "*distant*" supervision to help determine important features for the task. We propose a novel framework, called **DPE**, to integrate two kinds of mutually-complementing signals for synonym discovery, i.e., *distributional features* based on corpus-level statistics and *textual patterns* based on local contexts. In particular, DPE jointly optimizes the two kinds of signals in conjunction with distant supervision, so that they can mutually enhance each other in the training stage. At the inference stage, both signals will be utilized to discover synonyms for the given entities. Experimental results prove the effectiveness of the proposed framework.

PART II

Extracting Typed Relationships

CHAPTER 7

Joint Extraction of Typed Entities and Relationships

7.1 OVERVIEW AND MOTIVATION

The extraction of entities and their relations is critical to understanding massive text corpora. Identifying the token spans in text that constitute entity mentions and assigning types (e.g., person, company) to these spans as well as to the relations between entity mentions (e.g., employed_by) are key to structuring content from text corpora for further analytics. For example, when an extraction system finds a "produce" relation between "company" and "product" entities in news articles, it supports answering questions like *what products does company X produce?*." Once extracted, such structured information is used in many ways, e.g., as primitives in information extraction, KB population [Dong et al., 2014, West et al., 2014], and question-answering systems [Bian et al., 2008, Sun et al., 2015]. Traditional systems for relation extraction [Bach and Badaskar, 2007, Culotta and Sorensen, 2004, GuoDong et al., 2005] partition the process into several subtasks and solve them incrementally (i.e., detecting entities from text, labeling their types and then extracting their relations). Such systems treat the subtasks independently and so may propagate errors across subtasks in the process. Recent studies [Li and Ji, 2014, Miwa and Sasaki, 2014, Roth and Yih, 2007] focus on joint extraction methods to capture the inhereent linguistic dependencies between relations and entity arguments (e.g., the types of entity arguments help determine their relation type, and vice versa) to resolve error propagation.

A major challenge in joint extraction of typed entities and relations is to design *domain-independent* systems that will apply to text corpora from different domains in the *absence of human-annotated, domain data*. The process of manually labeling a training set with a large number of entity and relation types is too expensive and error-prone. The rapid emergence of large, domain-specific text corpora (e.g., news, scientific publications, social media content) calls for methods that can jointly extract entities and relations of target types with minimal or no human supervision.

Toward this goal, there are broadly two kinds of efforts: weak supervision and distant supervision. Weak supervision [Bunescu and Mooney, 2007, Etzioni et al., 2004, Nakashole et al., 2013] relies on a small set of manually specified seed instances (or patterns) that are applied in bootstrapping learning to identify more instances of each type. This assumes seeds are unambiguous and sufficiently frequent in the corpus, which requires careful seed selection by human [Bach and Badaskar, 2007]. Distant supervision [Hoffmann et al., 2011, Mintz et al., 2009, Riedel et al., 2010, Surdeanu et al., 2012] generates training data automatically by aligning texts and a KB (see Fig. 7.1). The typical workflow is: (1) detect entity mentions in text; (2) map detected entity mentions to entities in KB; (3) assign, to the candidate type set of each entity mention, all KB types of its KB-mapped entity; and (4) assign, to the candidate type set of each entity mention pair, all KB relation types between their KB-mapped entities. The automatically labeled training corpus is then used to infer types of the *remaining* candidate entity mentions and relation mentions (i.e., unlinkable candidate mentions).

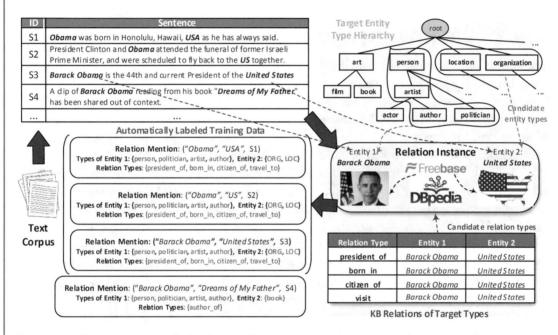

Figure 7.1: Current systems find relations (*Barack Obama, United States*) mentioned in sentences S1-S3 and assign the same relation types (entity types) to all relation mentions (entity mentions), when only some types are correct for context (highlighted in blue font).

In this chapter, we study the problem of *joint extraction of typed entities and relations with distant supervision*. Given a domain-specific corpus and a set of target entity and relation types from a KB, we aim to detect relation mentions (together with their entity arguments) from text, and categorize each in context by target types or Not-Target-Type (None), with distant supervision. Current distant supervision methods focus on solving the subtasks separately (e.g., extracting typed entities or relations), and encounter the following limitations when handling the joint extraction task.

• **Domain Restriction:** They rely on pre-trained named entity recognizers (or noun phrase chunker) to detect entity mentions. These tools are usually designed for a few general types (e.g., person, location, organization) and require additional human labors to work on specific domains (e.g., scientific publications).

• **Error Propagation:** In current extraction pipelines, incorrect entity types generated in entity recognition and typing step serve as features in the relation extraction step (i.e., errors are propagated from upstream components to downstream ones). Cross-task dependencies are ignored in most existing methods.

• **Label Noise:** In distant supervision, the context-agnostic mapping from relation (entity) mentions to KB relations (entities) may bring false positive type labels (i.e., label noise) into the automatically labeled training corpora and results in inaccurate models.

In Fig. 7.1, for example, all KB relations between entities *Barack Obama* and *United States* (e.g., born_in, president_of) are assigned to the relation mention in sentence S_1 (while only born_in is correct within the context). Similarly, all KB types for *Barack Obama* (e.g., politician, artist) are assigned to the mention "*Obama*" in S_1 (while only person is true). Label noise becomes an impediment to learn effective type classifiers. The larger the target type set, the more severe the degree of label noise (see Table 7.1).

Table 7.1: A study of type label noise. (1) %entity mentions with multiple *sibling entity types* (e.g., actor, singer) in the given entity type hierarchy; and (2) %relation mentions with multiple *relation types*, for the three experiment datasets.

Dataset	NYT [Riedel et al., 2010]	Wiki-KBP [Ellis et al,. 2014]	BioInfer [Pyysalo et al., 2007]
# of entity types	47	126	2,200
Noisy entity mentions (%)	20.32	28.31	59.80
# of relation types	24	19	94
Noisy relation mentions (%)	15.54	8.54	41.12

We approach the joint extraction task as follows. (1) Design a domain-agnostic text segmentation algorithm to detect candidate entity mentions with distant supervision and minimal linguistic assumption (i.e., assuming part-of-speech (POS) tagged corpus is given [Hovy et al., 2015]). (2) Model the mutual constraints between the types of the relation mentions and the types of their entity arguments, to enable feedbacks between the two subtasks. (3) Model the true type labels in a candidate type set as latent variables and require only the "*best*" type (progressively estimated as we learn the model) to be relevant to the mention—this is a less limiting requirement compared with existing multi-label classifiers that assume "*every*" candidate type is relevant to the mention.

To integrate these elements of our approach, a novel framework, CoType, is proposed. It first runs POS-constrained text segmentation using positive examples from KB to mine quality entity mentions, and forms candidate relation mentions (Section 7.3.1). Then CoType performs entity linking to map candidate relation (entity) mentions to KB relations (entities) and obtain the KB types. We formulate a global objective to *jointly* model (1) corpus-level co-occurrences between *linkable* relation (entity) mentions and text features extracted from their local contexts; (2) associations between mentions and their KB-mapped type labels; and (3) interactions between relation mentions and their entity arguments. In particular, we design a novel partial-label loss to model the noisy mention-label associations in a robust way, and adopt translation-based objective to capture the entity-relation interactions. Minimizing the objective yields two low-dimensional spaces (for entity and relation mentions, respectively), where, in each space, objects whose types are semantically close also have similar representation (see Section 7.3.2). With the learned embeddings, we can efficiently estimate the types for the remaining *unlinkable* relation mentions and their entity arguments (see Section 7.3.3).

The major contributions of this chapter are as follows.

1. A novel distant-supervision framework, CoTYPE, is proposed to extract typed entities and relations in domain-specific corpora with minimal linguistic assumption. (Fig. 7.2)

2. A domain-agnostic text segmentation algorithm is developed to detect entity mentions using distant supervision. (Section 7.3.1)

3. A joint embedding objective is formulated that models mention-type association, mention-feature co-occurrence, entity-relation cross-constraints in a noise-robust way. (Section 7.3.2)

4. Experiments with three public datasets demonstrate that CoTYPE improves the performance of state-of-the-art systems of entity typing and relation extraction significantly, demonstrating robust domain-independence. (Section 7.4)

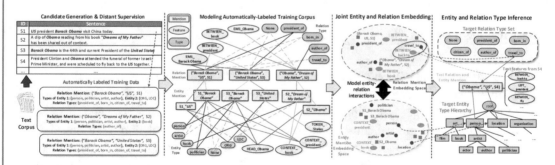

Figure 7.2: Framework overview of CoTYPE.

7.2   PRELIMINARIES

The input to our proposed CoTYPE framework is a POS-tagged text corpus $\mathcal{D}$, a KB Ψ (e.g., Freebase [Bollacker et al., 2008]), a target entity type hierarchy $\mathcal{Y}$, and a target relation type set $\mathbb{R}$. The target type set $\mathcal{Y}$ (set $\mathbb{R}$) covers a subset of entity (relation) types that the users are interested in from Ψ, i.e., $\mathcal{Y} \subset \mathcal{Y}_\Psi$ and $\mathbb{R} \subset \mathbb{R}_\Psi$.

Entity and Relation Mention. An *entity mention* (denoted by m) is a token span in text which represents an entity e. A *relation instance* $r(e_1, e_2, \ldots, e_n)$ denotes some type of relation $r \in \mathbb{R}$ between multiple entities. In this work, we focus on binary relations, i.e., $r(e_1, e_2)$. We define a *relation mention* (denoted by z) for some relation instance $r(e_1, e_2)$ as a (ordered) pair of entities mentions of e_1 and e_2 in a sentence s, and represent a relation mention with entity mentions m_1 and m_2 in sentence s as $z = (m_1, m_2, s)$.

Knowledge Bases and Target Types. A KB with a set of entities $\mathcal{E}_\Psi$ contains human-curated facts on both relation instances $\mathcal{I}_\Psi = \{r(e_1, e_2)\} \subset \mathbb{R}_\Psi \times \mathcal{E}_\Psi \times \mathcal{E}_\Psi$, and entity-type facts $\mathcal{T}_\Psi = \{(e, y)\} \subset \mathcal{E}_\Psi \times \mathcal{Y}_\Psi$. *Target entity type hierarchy* is a tree where nodes represent entity types of interests from the set $\mathcal{Y}_\Psi$. An entity mention may have multiple types, which together constitute one *type-path* (not required to end at a leaf) in the given type hierarchy. In existing studies, several entity type hierarchies are manually constructed using Freebase [Gillick et al., 2014, Lee et al., 2007] or WordNet [Yosef et al., 2012]. *Target relation type set* is a set of relation types of interests from the set $\mathbb{R}_\Psi$.

Automatically Labeled Training Data. Let $\mathcal{M} = \{m_i\}_{i=1}^N$ denote the set of entity mentions extracted from corpus $\mathcal{D}$. Distant supervision maps $\mathcal{M}$ to KB entities $\mathcal{E}_\Psi$ with an entity disambiguation system [Hoffart et al., 2011, Mendes et al., 2011] and heuristically assign type labels to the mapped mentions. In practice, only a small number of entity mentions in set $\mathcal{M}$ can be mapped to entities in $\mathcal{E}_\Psi$ (i.e., *linkable entity mentions*, denoted by $\mathcal{M}_L$). As reported in Lin et al. [2012] and Ren et al. [2015], the ratios of $\mathcal{M}_L$ over $\mathcal{M}$ are usually *lower than 50%* in domain-specific corpora.

Between any two linkable entity mentions m_1 and m_2 in a sentence, a relation mention z_i is formed if there exists one or more KB relations between their KB-mapped entities e_1 and e_2. Relations between e_1 and e_2 in KB are then associated with z_i to form its candidate relation type set $\mathbb{R}_i$, i.e., $\mathbb{R}_i = \{r \mid r(e_1, e_2) \in \mathbb{R}_\Psi\}$. In a similar way, types of e_1 and e_2 in KB are associated with m_1 and m_2 respectively, to form their candidate entity type sets $\mathcal{Y}_{i,1}$ and $\mathcal{Y}_{i,2}$, where $\mathcal{Y}_{i,x} = \{y \mid (e_x, y) \in \mathcal{Y}_\Psi\}$. Let $\mathcal{Z}_L = \{z_i\}_{i=1}^{N_L}$ denote the set of extracted relation mentions that can be mapped to KB. Formally, we represent the automatically labeled training corpus for the joint extraction task, denoted as $\mathcal{D}_L$, using a set of tuples $\mathcal{D}_L = \{(z_i, \mathbb{R}_i, \mathcal{Y}_{i,1}, \mathcal{Y}_{i,2})\}_{i=1}^{N_L}$.

Problem Description. By pairing up entity mentions (from set $\mathcal{M}$) within each sentence in $\mathcal{D}$, we generate a set of *candidate relation mentions*, denoted as $\mathcal{Z}$. Set $\mathcal{Z}$ consists of (1) linkable relation mentions $\mathcal{Z}_L$, (2) unlinkable (true) relation mentions, and (3) false relation mention (i.e., no target relation expressed between).

Let $\mathcal{Z}_U$ denote the set of *unlabeled* relation mentions in (2) and (3) (i.e., $\mathcal{Z}_U = \mathcal{Z} \setminus \mathcal{Z}_L$). Our main task is to determine the relation type label (from the set $\mathbb{R} \cup \{\text{None}\}$) for each relation mention in set $\mathcal{Z}_U$, and the entity type labels (either a single type-path in $\mathcal{Y}$ or None) for each entity mention argument in $z \in \mathcal{Z}_U$, using the automatically labeled corpus $\mathcal{D}_L$. Formally, we define the joint extraction of typed entities and relations task as follows.

Problem 7.1 **Joint Entity and Relation Extraction.** Given a POS-tagged corpus $\mathcal{D}$, a KB Ψ, a target entity type hierarchy $\mathcal{Y} \subset \mathcal{Y}_\Psi$ and a target relation type set $\mathbb{R} \subset \mathbb{R}_\Psi$, the joint extraction task aims to (1) detect entity mentions $\mathcal{M}$ from $\mathcal{D}$; (2) generate training data $\mathcal{D}_L$ with KB Ψ; and (3) estimate a relation type $r^* \in \mathbb{R} \cup \{\texttt{None}\}$ for each test relation mention $z \in \mathcal{Z}_U$ and a single type-path $\mathcal{Y}^* \subset \mathcal{Y}$ (or $\texttt{None}$) for each entity mention in z, using $\mathcal{D}_L$ and its context s.

Non-goals. This work relies on an entity linking system [Mendes et al., 2011] to provide disambiguation function, but we do not address their limits here (e.g., label noise introduced by wrongly mapped KB entities). We also assume human-curated target type hierarchies are given (It is out of the scope of this study to generate the type hierarchy).

7.3 THE COTYPE FRAMEWORK

This section lays out the proposed framework. The joint extraction task poses two unique challenges. First, type association in distant supervision between linkable entity (relation) mentions and their KB-mapped entities (relations) is *context-agnostic*—the candidate type sets $\{\mathbb{R}_i, \mathcal{Y}_{i,1}, \mathcal{Y}_{i,2}\}$ contain "false" types. Supervised learning [Gormley et al., 2015, GuoDong et al., 2005] may generate models biased to the incorrect type labels [Ren et al., 2016c]. Second, there exists dependencies between relation mentions and their entity arguments (e.g., type correlation). Current systems formulates the task as a *cascading* supervised learning problem and may suffer from error propagation.

Our solution casts the type prediction task as *weakly supervised learning* (to model the relatedness between mentions and their candidate types *in contexts*) and uses *relational learning* to capture interactions between relation mentions and their entity mention argument *jointly*, based on the redundant text signals in a large corpus.

Specifically, CoType leverages partial-label learning [Nguyen and Caruana, 2008] to faithfully model mention-type association using text features extracted from mentions' local contexts. It uses the translation embedding-based objective [Bordes et al., 2013] to model the mutual type dependencies between relation mentions and their entity (mention) arguments.

Framework Overview. We propose a *embedding-based* framework with distant supervision (see also Fig. 7.2) as follows.

1. Run POS-constrained text segmentation algorithm on POS-tagged corpus $\mathcal{D}$ using positive examples obtained from KB, to detect candidate entity mentions $\mathcal{M}$ (Section 7.3.1).

2. Generate candidate relation mentions $\mathcal{Z}$ from $\mathcal{M}$, extract text features for each relation mention $z \in \mathcal{Z}$ and their entity mention argument (Section 7.3.1). Apply distant supervision to generate labeled training data $\mathcal{D}_L$ (Section 7.2).

3. Jointly embed relation and entity mentions, text features, and type labels into two low-dimensional spaces (for entities and relations, respectively) where, in each space, close objects tend to share the same types (Section 7.3.2).

4. Estimate type labels r^* for each test relation mention $z \in \mathcal{Z}_U$ and type-path $\mathcal{Y}^*$ for each test entity mention m in $\mathcal{Z}_U$ from learned embeddings, by searching the target type set $\mathcal{Y}$ or the target type hierarchy $\mathbb{R}$ (Section 7.3.3).

7.3.1   CANDIDATE GENERATION

Entity Mention Detection. Traditional entity recognition systems [Finkel et al., 2005, Nadeau and Sekine, 2007] rely on a set of linguistic features (e.g., dependency parse structures of a sentence) to train sequence labeling models (for a few common entity types). However, sequence labeling models trained on automatically labeled corpus $\mathcal{D}_L$ may not be effective, as distant supervision only annotates a small number of entity mentions in $\mathcal{D}_L$ (thus generates a lot of "false negative" token tags). To address domain restriction, we develop a distantly supervised text segmentation algorithm for *domain-agnostic entity detection*. By using quality examples from KB as guidance, it partitions sentences into segments of entity mentions and words, by incorporating (1) corpus-level concordance statistics; (2) sentence-level lexical signals; and (3) grammatical constraints (i.e., POS tag patterns).

We extend the methodology used in El-Kishky et al. [2014] and Liu et al. [2015], to model the *segment quality* (i.e., "how likely a candidate segment is an entity mention") as a combination of *phrase quality* and *POS pattern quality*, and use positive examples in $\mathcal{D}_L$ to estimate the segment quality. The workflow is as follows: (1) mine frequent contiguous patterns for both word sequence and POS tag sequence up to a fixed length from POS-tagged corpus $\mathcal{D}$; (2) extract features including corpus-level concordance and sentence-level lexical signals to train two random forest classifiers [Liu et al., 2015], for estimating quality of candidate phrase and candidate POS pattern; (3) find the best segmentation of $\mathcal{D}$ using the estimated segment quality scores (see Eq. (7.1)); and (4) compute rectified features using the segmented corpus and repeat steps (2)–(4) until the result converges:

$$p\big(b_{t+1}, c \mid b_t\big) = p\big(b_{t+1} - b_t\big) \cdot p\big(c \mid b_{t+1} - b_t\big) \cdot Q(c). \qquad (7.1)$$

Specifically, we find the best segmentation S_d for each document d (in $\mathcal{D}$) by maximizing the "joint segmentation quality", defined as $\sum_d^{\mathcal{D}} \log p(S_d, d) = \sum_d^{\mathcal{D}} \sum_{t=1}^{|d|} \log p(b_{t+1}^{(d)}, c^{(d)} \mid b_t^{(d)})$, where $p(b_{t+1}^{(d)}, c^{(d)} \mid b_t^{(d)})$ denote the probability that segment $c^{(c)}$ (with starting index $b_{t+1}^{(d)}$ and ending index in document d) is a good entity mention, as defined in Eq. (7.1). The first term in Eq. (7.1) is a segment length prior, the second term measures how likely segment c is generated given a length $(b_{t+1} - b_t)$ (to be estimated), and the third term denotes the segment quality. In this work, we define function $Q(c)$ as the *equally weighted combination of the phrase quality score and POS pattern quality score* for candidate segment c, which is estimated in step (2). The joint probability can be efficiently maximize using Viterbi Training with time complexity linear to the corpus size [Liu et al., 2015]. The segmentation result provides us a set of candidate entity mentions, forming the set $\mathcal{M}$.

Table 7.2 compares our entity detection module with a sequence labeling model [Ling and Weld, 2012] (linear-chain CRF) trained on the labeled corpus $\mathcal{D}_L$ in terms of F1 score. Figure 7.3 show the high-/low-quality POS patterns learned using entity names found in $\mathcal{D}_L$ as examples.

	POS Tag Pattern	Example
Good (high score)	NNP NNP NN NN CD NN JJ NN	San Francisco/Barack Obama/United States comedy drama/car accident/club captain seven network/seven dwarfs/2001 census crude oil/nucletic acid/baptist church
Bad (low score)	DT JJ NND CD CD NN IN NN IN NNP NNP VVD RB IN	a few miles/the early stages/the late 1980s 2 : 0 victory over/1 : 0 win over rating on rotten tomatoes worked together on/spent much of

Figure 7.3: Example POS tag patterns learned using KB examples.

Table 7.2: Comparison of F1 scores on entity mention detection

Dataset	NYT	Wiki-KBP	BioInfer
FIGER segmenter [Ling and Weld, 2012]	0.751	0.814	0.652
Our Approach	**0.837**	**0.833**	**0.785**

Relation Mention Generation. We follow the procedure introduced in Section 7.2 to generate the set of candidate relation mentions $\mathcal{Z}$ from the detected candidate entity mentions $\mathcal{M}$: for each pair of entity mentions (m_a, m_b) found in sentence s, we form two candidate relation mentions $z_1 = (m_a, m_b, s)$ and $z_2 = (m_b, m_a, s)$. Distant supervision is then applied on $\mathcal{Z}$ to generate the set of KB-mapped relation mentions $\mathcal{Z}_L$. Similar to Hoffmann et al. [2011] and Mintz et al. [2009], we sample 30% unlinkable relation mentions between two KB-mapped entity mentions (from set $\mathcal{M}_L$) in a sentence as examples for modeling None relation label, and sample 30% unlinkable entity mentions (from set $\mathcal{M} \setminus \mathcal{M}_L$) to model None entity label. These negative examples, together with type labels for mentions in $\mathcal{Z}_L$, form the automatically labeled data $\mathcal{D}_L$ for the task.

Text Feature Extraction. To capture the shallow syntax and distributional semantics of a relation (or entity) mention, we extract various lexical features from both mention itself (e.g., head token) and its context s (e.g., bigram), in the POS-tagged corpus. Table 7.3 lists the set of text features for relation mention, which is similar to those used in Chan and Roth [2010] and Mintz et al. [2009] (excluding the dependency parse-based features and entity type features). We use the same set of features for entity mentions as those used in Ling and Weld [2012] and Ren et al. [2016c]. We denote the set of M_z (M_m) unique features extracted of relation mentions $\mathcal{Z}_L$ (entity mentions in $\mathcal{Z}_L$) as $\mathcal{F}_z = \{f_j\}_{j=1}^{M_z}$ (and $\mathcal{F}_m = \{f_j\}_{j=1}^{M_m}$).

Table 7.3: Text features for relation mentions used in this work [GuoDong et al., 2005, Riedel et al., 2010] (excluding dependency parse-based features and entity type features). ("*Barack Obama*," "*United States*") is used as an example relation mention from the sentence "*Honolulu native* **Barack Obama** *was elected President of the* **United States** *on March 20 in 2008.*"

Feature	Description	Example
Entity mention (EM) head	Syntactic head token of each entity mention	"*HEAD_EM1_Obama*"
Entity Mention Token	Tokens in each entity mention	"*TKN_EM1_Barack*"
Tokens between two EMs	Each token between two EMs	"*was*", "*elected*", "*President*", "*of*", "*the*"
Part-of-speech (POS) tag	POS tags of tokens between two EMs	"*VBD*", "*VBN*", "*NNP*", "*IN*", "*DT*"
Collocations	Bigrams in left/right 3-word window of each EM	"*Honolulu native*", "*native Barack*", ...
Entity mention order	Whether EM 1 is before EM 2	"*EM1_BEFORE_EM2*"
Entity mention distance	Number of tokens between the two EMs	"*EM_DISTANCE_5*"
Entity mention context	Unigrams before and after each EM	"*native*", "*was*", "*the*", "*in*"
Special pattern	Occurrence of pattern "em1_in_ em2"	"*PATTERN_NULL*"
Brown cluster (learned on $\mathcal{D}$)	Brown cluster ID for each token	"*8_1101111*", "*12_111011111111*"

7.3.2   JOINT ENTITY AND RELATION EMBEDDING

This section formulates a joint optimization problem for embedding different kinds of interactions between linkable relation mentions $\mathcal{Z}_L$, linkable entity mentions $\mathcal{M}_L$, entity and relation type labels $\{\mathbb{R}, \mathcal{Y}\}$ and text features $\{\mathcal{F}_z, \mathcal{F}_m\}$ into a d-dimensional *relation vector space* and a d-dimensional *entity vector space*. In each space, objects whose types are close to each other should have similar representation (e.g., see the third column in Fig. 7.2).

As the extracted objects and the interactions between them form a heterogeneous graph (see the second column in Fig. 7.2), a simple solution is to embed the whole graph into a *single* low-dimensional space [He and Niyogi, 2004, Ren et al., 2015]. However, such a solution encounters several problems: (1) false types in candidate type sets (i.e., false mention-type links in the graph) negatively impact the ability of the model to determine mention's true types; and (2) a single embedding space cannot capture the differences in entity and relation types (i.e., strong link between a relation mention and its entity mention argument does not imply that they have similar types).

In our solution, we propose a novel global objective, which extends a margin-based rank loss [Nguyen and Caruana, 2008] to model *noisy mention-type associations* and leverages the second-order proximity idea [Tang et al., 2015] to model corpus-level *mention-feature co-occurrences*. In particular, to capture the *entity-relation interactions*, we adopt a translation-based embedding loss [Bordes et al., 2013] to bridge the vector spaces of entity mentions and relation mentions.

Modeling Types of Relation Mentions. We consider both *mention-feature co-occurrences* and *mention-type associations* in the modeling of relation types for relation mentions in set $\mathcal{Z}_L$.

Intuitively, two relation mentions sharing many text features (i.e., with similar distribution over the set of text features $\mathcal{F}_m$) likely have similar relation types; and text features co-occurring with many relation mentions in the corpus tend to represent close type semantics. We propose the following hypothesis to guide our modeling of corpus-level mention-feature co-occurrences.

> **Hypothesis 7.1: Mention-Feature Co-occurrence**
>
> Two entity mentions tend to share similar types (close to each other in the embedding space) if they share many text features in the corpus, and the converse way also holds.

For example, in column 2 of Fig. 7.2, (*"Barack Obama,"* *"US,"* S_1) and (*"Barack Obama,"* *"United States,"* S_3) share multiple features including context word *"president"* and first entity mention argument *"Barack Obama,"* and thus they are likely of the same relation type (i.e., president_of).

Formally, let vectors $\mathbf{t}z_i$, $\mathbf{t}c_j \in \mathbb{R}^d$ represent relation mention $z_i \in \mathcal{Z}_L$ and text feature $f_j \in \mathcal{F}_z$ in the d-dimensional *relation embedding space*. Similar to the distributional hypothesis [Mikolov et al., 2013] in text corpora, we apply second-order proximity [Tang et al., 2015] to model the idea that *objects with similar distribution over neighbors are similar to each other* as follows:

$$\mathcal{L}_{ZF} = -\sum_{z_i \in \mathcal{Z}_L} \sum_{f_j \in \mathcal{F}_z} w_{ij} \cdot \log p(f_j | z_i), \qquad (7.2)$$

where $p(f_j|z_i) = \exp(tz_i^T tc_j)/\sum_{f' \in \mathcal{F}_z} \exp(tz_i^T tc_{j'})$ denotes the probability of f_j generated by z_i, and w_{ij} is the co-occurrence frequency between (z_i, f_j) in corpus $\mathcal{D}$. Function $\mathcal{L}_{ZF}$ in Eq. (7.2) enforces the conditional probability specified by embeddings, i.e., $p(\cdot|z_i)$ to be close to the empirical distribution.

To perform efficient optimization by avoiding summation over all features, we adopt negative sampling strategy [Mikolov et al., 2013] to sample multiple *false* features for each (z_i, f_j), according to some *noise distribution* $P_n(f) \propto D_f^{3/4}$ [Mikolov et al., 2013] (with D_f denotes the number of relation mentions co-occurring with f). Term $\log p(f_j|z_i)$ in Eq. (7.2) is replaced with the term as follows:

$$\log \sigma \left(tz_i^T tc_j \right) + \sum_{v=1}^{V} \mathbb{E}_{f_{j'} \sim P_n(f)} \left[\log \sigma \left(-tz_i^T tc_{j'} \right) \right], \tag{7.3}$$

where $\sigma(x) = 1/(1 + \exp(-x))$ is the sigmoid function. The first term in Eq. (7.3) models the observed co-occurrence, and the second term models the Z negative feature samples.

In D_L, each relation mention z_i is heuristically associated with a set of candidate types $\mathbb{R}_i$. Existing embedding methods rely on either the *local consistent assumption* [He and Niyogi, 2004] (i.e., objects strongly connected tend to be similar) or the *distributional assumption* [Mikolov et al., 2013] (i.e., objects sharing similar neighbors tend to be similar) to model object associations. However, some associations between z_i and $r \in \mathbb{R}_i$ are "false" associations and adopting the above assumptions may incorrectly yield mentions of different types having similar vector representations. For example, in Fig. 7.1, mentions ("*Obama*," "*U.S.*," S_1) and ("*Obama*," "*U.S.*," S_2) have several candidate types in common (thus high distributional similarity), but their true types are different (i.e., born_in vs. travel_to).

We specify the likelihood of "*whether the association between a relation mention and its candidate entity type being true*" as the *relevance* between these two kinds of objects (measured by the similarity between their current estimated embedding vectors). To impose such idea, we model the associations between each linkable relation mention z_i (in set $\mathcal{Z}_L$) and its noisy candidate relation type set $\mathbb{R}_i$ based on the following hypothesis.

Hypothesis 7.2: Partial-Label Association

A relation mention's embedding vector should be more similar (closer in the low-dimensional space) to its "most relevant" candidate type, than to any other non-candidate type.

Specifically, we use vector $tr_k \in \mathbb{R}^d$ to represent relation type $r_k \in \mathbb{R}$ in the embedding space. The similarity between (z_i, r_k) is defined as the dot product of their embedding vectors, i.e., $\phi(z_i, r_k) = \mathbf{t}z_i^T \mathbf{t}r_k$. We extend the margin-based loss in Nguyen and Caruana [2008] and define a partial-label loss ℓ_i for each relation mention $z_i \in \mathcal{M}_L$ as follows:

$$\ell_i = \max \left\{ 0, 1 - \left[\max_{r \in \mathbb{R}_i} \phi(z_i, r) - \max_{r' \in \overline{\mathcal{R}}_i} \phi(z_i, r') \right] \right\}. \tag{7.4}$$

The intuition behind Eq. (7.4) is that: for relation mention z_i, the maximum similarity score associated with its candidate type set $\mathbb{R}_i$ should be greater than the maximum similarity score associated with any other *non-candidate types* $\overline{\mathcal{R}}_i = \mathbb{R} \setminus \mathbb{R}_i$. Minimizing ℓ_i forces z_i to be embedded closer to the *most "relevant"* type in $\mathbb{R}_i$, than to any other non-candidate types in $\overline{\mathcal{R}}_i$. This contrasts sharply with multi-label learning [Ling and Weld, 2012], where m_i is embedded closer to *every* candidate type than any other non-candidate type.

To faithfully model the types of relation mentions, we integrate the modeling of mention-feature co-occurrences and mention-type associations by the following objective:

$$O_Z = \mathcal{L}_{ZF} + \sum_{i=1}^{N_L} \ell_i + \frac{\lambda}{2} \sum_{i=1}^{N_L} \|\mathbf{t}z_i\|_2^2 + \frac{\lambda}{2} \sum_{k=1}^{K_r} \|\mathbf{t}r_k\|_2^2, \tag{7.5}$$

where tuning parameter $\lambda > 0$ on the regularization terms is used to control the scale of the embedding vectors.

By doing so, text features, as complements to mention's candidate types, also participate in modeling the relation mention embeddings, and help identify a mention's most relevant type—mention-type relevance is progressively estimated during model learning. For example, in the left column of Fig. 7.4, context words "*president*" helps infer that relation type `president_of` is more relevant (i.e., higher similarity between the embedding vectors) to relation mention ("*Mr. Obama*," "*U.S.*," S_2), than type `born_in` does.

Modeling Types of Entity Mentions. In a way similar to the modeling of types for relation mentions, we follow Hypotheses 7.1 and 7.2 to model types of entity mentions. In Fig. 7.2 (col. 2), for example, entity mentions "S_1_*Barack Obama*" and "S_3_*Barack Obama*" share multiple text features in the corpus, including head token "*Obama*" and context word "*president*," and thus tend to share the same entity types like `politician` and `person` (i.e., Hypothesis 7.1). Meanwhile, entity mentions "S_1_*Barack Obama*" and "S_2_*Obama*" have the same candidate entity types but share very few text features in common. This implies that likely their true type labels are different. Relevance between entity mentions and their true type labels should be progressively estimated based on the text features extracted from their local contexts (i.e., Hypothesis 7.2).

Formally, let vectors $\mathbf{t}m_i, \mathbf{t}c_j', \mathbf{t}y_k \in \mathbb{R}^d$ represent entity mention $m_i \in \mathcal{M}_L$, text features (for entity mentions) $f_j \in \mathcal{F}_m$, and entity type $y_k \in \mathcal{Y}$ in a d-dimensional *entity embedding space*, respectively. We model the corpus-level co-occurrences between entity mentions and text features by second-order proximity as follows:

$$\mathcal{L}_{MF} = - \sum_{m_i \in \mathcal{M}_L} \sum_{f_j \in \mathcal{F}_m} w_{ij} \cdot \log p(f_j | m_i), \tag{7.6}$$

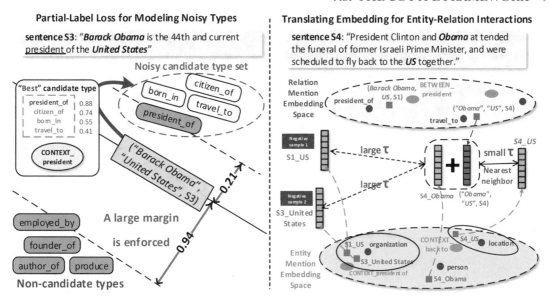

Figure 7.4: Illustrations of the partial-label associations, Hypothesis 7.2 (the left col.), and the entity-relation interactions, Hypothesis 7.3 (the right col.).

where the conditional probability term $\log p(f_j|m_i)$ is defined as $\log p(f_j|m_i) = \log \sigma(tm_i^T tc'_j) + \sum_{v=1}^{V} \mathbb{E}_{f_{j'} \sim P_n(f)}[\log \sigma(-tm_i^T tc'_{j'})]$. By integrating the term $\mathcal{L}_{MF}$ with partial-label loss $\ell'_i = \max\{0, 1 - [\max_{y \in \mathcal{Y}_i} \phi(m_i, y) - \max_{y' \in \overline{\mathcal{Y}}_i} \phi(m_i, y')]\}$ for N'_L unique linkable entity mentions (in set $\mathcal{M}_L$), we define the objective function for modeling types of entity mentions as follows.

$$O_M = \mathcal{L}_{MF} + \sum_{i=1}^{N'_L} \ell'_i + \frac{\lambda}{2} \sum_{i=1}^{N'_L} \|tm_i\|_2^2 + \frac{\lambda}{2} \sum_{k=1}^{K_y} \|ty_k\|_2^2. \qquad (7.7)$$

Minimizing the objective O_M yields an entity embedding space where, in that space, objects (e.g., entity mentions, text features) close to each other will have similar types.

Modeling Entity-Relation Interactions. In reality, there exists different kinds of interactions between a relation mention $z = (m_1, m_2, s)$ and its entity mention arguments m_1 and m_2. One major kind of interactions is the correlation between relation and entity types of these objects— entity types of the two entity mentions provide good hints for determining the relation type of the relation mention, and vice versa. For example, in Fig. 7.4 (right column), knowing that entity mention "S_4_US" is of type location (instead of organization) helps determine that relation mention ("Obama," "U.S.," S_4) is more likely of relation type travel_to, rather than relation types like president_of or citizen_of.

Intuitively, entity types of the entity mention arguments pose constraints on the search space for the relation types of the relation mention (e.g., it is unlikely to find a author_of relation between a `organization` entity and a `location` entity). The proposed Hypotheses 7.1 and 7.2 model types of relation mentions and entity mentions by learning an entity embedding space and a relation embedding space, respectively. The correlations between entity and relation types (and their embedding spaces) motivates us to model entity-relation interactions based on the following hypothesis.

Hypothesis 7.3: Entity-Relation Interaction

For a relation mention $z = (m_1, m_2, s)$, embedding vector of m_1 should be a nearest neighbor of the embedding vector of m_2 plus the embedding vector of relation mention z.

Given the embedding vectors of any two members in $\{z, m_1, m_2\}$, say tz and tm_1, Hypothesis 7.3 forces the "$tm_1 + tz \approx tm_2$." This helps regularize the learning of vector tm_2 (which represents the type semantics of entity mention m_2) in addition to the information encoded by objective O_M in Eq. (7.7). Such a "translating operation" between embedding vectors in a low-dimensional space has been proven effective in embedding entities and relations in a structured knowledge baes [Bordes et al., 2013]. We extend this idea to model the type correlations (and mutual constraints) between embedding vectors of entity mentions and embedding vectors of relation mentions, which are modeled in two different low-dimensional spaces.

Specifically, we define error function for the triple of a relation mention and its two entity mention arguments (z, m_1, m_2) using ℓ-2 norm: $\tau(z) = \|tm_1 + tz - tm_2\|_2^2$. A small value on $\tau(z)$ indicates that the embedding vectors of (z, m_1, m_2) do capture the type constraints. To enforce small errors between linkable relation mentions (in set $\mathcal{Z}_L$) and their entity mention arguments, we use margin-based loss [Bordes et al., 2013] to formulate a objective function as follows:

$$O_{ZM} = \sum_{z_i \in \mathcal{Z}_L} \sum_{v=1}^{V} \max\left\{0, 1 + \tau(z_i) - \tau(z_v)\right\}, \tag{7.8}$$

where $\{z_v\}_{v=1}^{V}$ are negative samples for z, i.e., z_v is randomly sampled from the negative sample set $\{(z', m_1, m_2)\} \cup \{(z, m_1', m_2)\} \cup \{(z, m_1, m_2')\}$ with $z' \in \mathcal{Z}_L$ and $m' \in \mathcal{M}_L$ [Bordes et al., 2013]. The intuition behind Eq. (7.8) is simple (see also the right col. in Fig. 7.4): embedding vectors for a relation mention and its entity mentions are modeled in the way that, the translating error τ between them should be *smaller* than the translating error of any negative sample.

A Joint Optimization Problem. Our goal is to embed all the available information for relation and entity mentions, relation, and entity type labels, and text features into a d-dimensional entity space and a d-dimensional relation space, following the three proposed hypotheses. An intuitive

Table 7.4: Notations

$\mathcal{D}$	Automatically generated training corpus
$\mathcal{M} = \{m_i\}_{i=1}^N$	Entity mentions in $\mathcal{D}$ (size N)
$\mathcal{Y} = \{y_k\}_{k=1}^K$	Target entity types (size K)
$\mathcal{Y}_i$	Candidate types of m_i
$\bar{\mathcal{Y}}_i$	Non-candidate types of m_i, i.e., $\bar{\mathcal{Y}}_i = \mathcal{Y} \setminus \mathcal{Y}_i$
$\mathcal{F} = \{f_j\}_{j=1}^M$	Text features in $\mathcal{D}$ (size M)
$\mathbf{u}_i \in \mathbb{R}^d$	Embedding of mention m_i (dim. d)
$\mathbf{c}_j \in \mathbb{R}^d$	Embedding of feature f_j (dim. d)
$\mathbf{v}_k, \mathbf{v}_k' \in \mathbb{R}^d$	Embeddings of type y_k on two views (dim. d)

solution is to *collectively* minimize the three objectives $\mathcal{O}_Z$, $\mathcal{O}_M$, and $\mathcal{O}_{ZM}$, as the embedding vectors of entity and relation mentions are shared across them. To achieve the goal, we formulate a joint optimization problem as follows:

$$\min_{\{tz_i\},\{tc_j\},\{tr_k\},\{tm_i\},\{tc_j'\},\{ty_k\}} \mathcal{O} = \mathcal{O}_M + \mathcal{O}_Z + \mathcal{O}_{ZM}. \tag{7.9}$$

Optimizing the global objective O in Eq. (7.9) enables the learning of entity and relation embeddings to be *mutually* influenced, such that, errors in each component can be constrained and corrected by the other. The joint embedding learning also helps the algorithm to find the true types for each mention, besides using text features.

In Eq. (7.9), one can also minimize the weighted combination of the three objectives $\{O_Z, O_M, O_{ZM}\}$ to model the importance of different signals, where weights could be manually determined or automatically learned from data. We leave this as future work.

7.3.3   MODEL LEARNING AND TYPE INFERENCE

The joint optimization problem in Eq. (7.9) can be solved in multiple ways. One solution is to first learn entity mention embeddings by minimizing $\mathcal{O}_M$, then apply the learned embeddings to optimize $\mathcal{O}_{MZ} + \mathcal{O}_Z$. However, such a solution does not fully exploit the entity-relation inter-actions in providing mutual feedbacks between the learning of entity mention embeddings and the learning of relation mention embeddings (see CoType-TwoStep in Section 7.4).

We design a stochastic sub-gradient descent algorithm [Shalev-Shwartz et al., 2011] based on edge sampling strategy [Tang et al., 2015], to efficiently solve Eq. (7.9). In each iter-ation, we alternatively sample from each of the three objectives $\{O_Z, O_M, O_{ZM}\}$ a batch of edges (e.g., (z_i, f_j)) and their negative samples, and update each embedding vector based on the derivatives. The proof procedure in Shalev-Shwartz et al. [2011] can be adopted to prove convergence of the proposed algorithm (to the local minimum). Equation (7.9) can also be solved by a mini-batch extension of the Pegasos algorithm [Shalev-Shwartz et al., 2011], which is a stochastic sub-gradient descent method and thus can efficiently handle massive text corpora. Due to lack of space, we do not include derivation details here.

Type Inference. With the learned embeddings of features and types in relation space (i.e., $\{tc_i\}$, $\{tr_k\}$) and entity space (i.e., $\{tc'_i\}$, $\{ty_k\}$), we can perform nearest neighbor search in the target relation type set $\mathbb{R}$, or a top-down search on the target entity type hierarchy $\mathcal{Y}$, to estimate the relation type (or the entity type-path) for each (unlinkable) test relation mention $z \in \mathcal{Z}_U$ (test entity mention $m \in \mathcal{M} \setminus \mathcal{M}_L$). Specifically, on the entity type hierarchy, we start from the tree's root and recursively find the best type among the children types by measuring the *cosine similarity* between entity type embedding and the vector representation of m in our learned entity embedding space. By extracting text features from m's local context (denoted by set $\mathcal{F}_m(m)$), we represent m in the learned entity embedding space using the vector $tm = \sum_{f_j \in \mathcal{F}_m(m)} tc'_j$. Similarly, for test relation mention z, we represent it in our learned relation embedding space by $tz = \sum_{f_j \in \mathcal{F}_z(z)} tc_j$ where $\mathcal{F}_z(z)$ is the set of text features extracted from z's local context s. The search process stops when we reach to a leaf type on the type hierarchy, or the similarity score is below a pre-defined threshold $\eta > 0$. If the search process returns an empty type-path (or type set), we output the predicted type label as None for the mention.

Computational Complexity Analysis. Let E be the total number of objects in CoType (entity and relation mentions, text features and type labels). By alias table method [Tang et al., 2015], setting up alias tables takes $O(E)$ time for all the objects, and sampling a negative example takes constant time. In each iteration of the CoType algorithm, optimization with negative sampling (i.e., optimizing second-order proximity and translating objective) takes $O(dV)$, and optimization with partial-label loss takes $O(dV(|\mathbb{R}| + |\mathcal{Y}|))$ time. Similar to Tang et al. [2015], we find the number of iterations for the algorithm to converge is usually *proportional to* the number of object interactions extracted from $\mathcal{D}$ (e.g., unique mention-feature pairs and mention-type associations), denoted as R. Therefore, the overall time complexity of CoType is $O(dRV(|\mathbb{R}| + |\mathcal{Y}|))$ (as $R \geq E$), which is *linear* to the total number of object interactions R in the corpus.

7.4 EXPERIMENTS

7.4.1 DATA PREPARATION AND EXPERIMENT SETTING

Our experiments use three public datasets[1] from different domains. (1) **NYT** [Riedel et al., 2010]: The training corpus consists of 1.18M sentences sampled from ~294K 1987–2007 New York Times news articles. 395 sentences are manually annotated by Hoffmann et al. [2011] to form the test data; (2) **Wiki-KBP** [Ling and Weld, 2012]: It uses 1.5M sentences sampled from ~780K Wikipedia articles [Ling and Weld, 2012] as training corpus and 14K manually

---

[1]Codes and datasets used in this chapter can be downloaded at: `https://github.com/shanzhenren/CoType`.

annotated sentences from 2013 KBP slot filling assessment results [Ellis et al., 2014] as test data; and (3) **BioInfer** [Pyysalo et al., 2007]: It consists of 1,530 manually annotated biomedical paper abstracts as test data and 100K sampled PubMed paper abstracts as training corpus. Statistics of the datasets are shown in Table 7.5.

Table 7.5: Statistics of the datasets in our experiments

Dataset	NYT	Wiki-KBP	BioInfer
# Relation/entity types	24/47	19/126	94/2,200
# Documents (in $\mathcal{D}$)	294,977	780,549	101,530
# Sentences (in $\mathcal{D}$)	1.18 M	1.51 M	521 k
# Training RMs (in $\mathcal{Z}_L$)	353 k	148 k	28 k
# Training EMs (in $\mathcal{Z}_L$)	701 k	247 k	53 k
# Text features (from $\mathcal{D}_L$)	2.6 M	1.3 M	575 k
# Text sentences (from $\mathcal{Z}_U$)	395	448	708
# Ground-truth RMs	3,880	2,948	3,859
# Ground-truth EMs	1,361	1,285	2,389

Automatically Labeled Training Corpora. The NYT training corpus has been heuristically labeled using distant supervision following the procedure in Riedel et al. [2010]. For Wiki-KBP and BioInfer training corpora, we utilized DBpedia Spotlight,[2] a state-of-the-art entity disambiguation tool, to map the detected entity mentions $\mathcal{M}$ to Freebase entities. We then followed the procedure introduced in Sections 7.2 and 7.3.1 to obtain candidate entity and relation types, and constructed the training data $\mathcal{D}_L$. For target types, we discard the relation/entity types which cannot be mapped to Freebase from the test data while keeping the Freebase entity/relation types (not found in test data) in the training data (see Table 7.5 for the type statistics).

Feature Generation. Table 7.3 lists the set of text features of relation mentions used in our experiments. We followed Ling and Weld [2012] to generate text features for entity mentions. Dependency parse-based features were excluded as only POS-tagged corpus is given as input. We used a six-word window to extract context features for each mention (three words on the left and the right). We applied the Stanford CoreNLP tool [Manning et al., 2014] to get POS tags. Brown clusters were derived for each corpus using public implementation.[3] The same kinds of features were used in all the compared methods in our experiments.

Evaluation Sets. For all three datasets, we used the provided training/test set partitions of the corpora. In each dataset, relation mentions in sentences are manually annotated with their relation types and the entity mention arguments are labeled with entity type-paths (see Table 7.5

[2]http://spotlight.dbpedia.org/
[3]https://github.com/percyliang/brown-cluster

for the statistics of test data). We further created a *validation set* by randomly sampling 10% mentions from each test set and used the remaining part to form the *evaluation set*.

Compared Methods. We compared CoType with its variants which model parts of the proposed hypotheses. Several state-of-the-art relation extraction methods (e.g., supervised, embedding, neural network) were also implemented (or tested using their published codes): (1) **DS+Perceptron** [Ling and Weld, 2012]: adopts multi-label learning on automatically labeled training data $\mathcal{D}_L$; (2) **DS+Kernel** [Mooney and Bunescu, 2005]: applies bag-of-feature kernel [Mooney and Bunescu, 2005] to train a SVM classifier using $\mathcal{D}_L$; (3) **DS+Logistic** [Mintz et al., 2009]: trains a multi-class logistic classifier[4] on $\mathcal{D}_L$; (4) **DeepWalk** [Perozzi et al., 2014]: embeds mention-feature co-occurrences and mention-type associations as a homogeneous network (with binary edges); (5) **LINE** [Tang et al., 2015]: uses second-order proximity model with edge sampling on a feature-type bipartite graph (where edge weight w_{jk} is the number of relation mentions having feature f_j and type r_k); (6) **MultiR** [Hoffmann et al., 2011]: is a state-of-the-art distant supervision method, which models noisy label in $\mathcal{D}_L$ by multi-instance multi-label learning; (7) **FCM** [Gormley et al., 2015]: adopts neural language model to perform compositional embedding; and (8) **DS-Joint** [Li and Ji, 2014]: jointly extract entity and relation mentions using structured perceptron on human-annotated sentences. We used $\mathcal{D}_L$ to train the model.

For CoType, besides the proposed model, **CoType**, we compare: (1) **CoType-RM**: This variant only optimize objective O_Z to learning feature and type embeddings for relation mentions; and (2) **CoType-TwoStep**: It first optimizes $\mathcal{O}_M$, then use the learned entity mention embedding $\{tm_i\}$ to initialize the minimization of $O_Z + O_{ZM}$—it represents a "pipeline" extraction diagram.

To test the performance on entity recognition and typing, we also compare with several entity recognition systems, including a supervised method **HYENA** [Yosef et al., 2012], distant supervision methods (**FIGER** [Ling and Weld, 2012], **Google** [Gillick et al., 2014], **WSABIE** [Yogatama et al., 2015]), and a noise-robust approach **PLE** [Ren et al., 2016c].

Parameter Settings. In our testing of CoType and its variants, we set $\alpha = 0.025$, $\eta = 0.35$ and $\lambda = 10^{-4}$ based on the analysis on validation sets. For convergence criterion, we stopped the loop in the algorithm if the relative change of $\mathcal{O}$ in Eq. (7.9) is smaller than 10^{-4}. For fair comparison, the dimensionality of embeddings d was set to 50 and the number of negative samples V was set to 5 for all embedding methods, as used in Tang et al. [2015]. For other tuning parameters in the compared methods, we tuned them on validation sets and picked the values which lead to the best performance.

Evaluation Metrics. For entity recognition and typing, we to use strict, micro, and macro F1 scores, as used in Ling and Weld [2012], for evaluating both detected entity mention boundaries and predicted entity types. We consider two settings in evaluation of relation extraction. For relation classification, ground-truth relation mentions are given and None label is excluded. We focus on testing type classification accuracy. For relation extraction, we adopt standard Precision

---

[4]We use liblinear package from https://github.com/cjlin1/liblinear

(**P**), Recall (**R**), and F1 score [Bach and Badaskar, 2007, Mooney and Bunescu, 2005]. Note that all our evaluations are *sentence-level* (i.e., context-dependent), as discussed in Hoffmann et al. [2011].

7.4.2 EXPERIMENTS AND PERFORMANCE STUDY

1. Performance on Entity Recognition and Typing. Among the compared methods, only FIGER [Ling and Weld, 2012] can detect entity mention. We apply our detection results (i.e., $\mathcal{M}$) as input for other methods. Table 7.6 summarizes the comparison results on the three datasets. Overall, CoTYPE outperforms others on all metrics on all three datasets (e.g., it obtains a 8% improvement on Micro-F1 over the *next best method* on NYT dataset). Such performance gains mainly come from: (1) a more robust way of modeling noisy candidate types (as compared to supervised method and distant supervision methods which ignore label noise issue); and (2) the joint embedding of entity and relation mentions in a mutually enhancing way (vs. the noise-robust method PLE [Ren et al., 2016c]). This demonstrates the effectiveness of enforcing Hypothesis 7.3 in CoTYPE framework.

Table 7.6: Performance comparison of entity recognition and typing (using strict, micro and macro metrics [Ling and Weld, 2012]) on the three datasets

Method	NYT			Wiki-KBP			BioInfer		
	S-F1	Ma-F1	Mi-F1	S-F1	Ma-F1	Mi-F1	S-F1	Ma-F1	Mi-F1
FIGER [Ling and Weld, 2012]	0.40	0.51	0.46	0.29	0.56	0.54	0.69	0.71	0.71
Google [Gillick et al., 2014]	0.38	0.57	0.52	0.30	0.50	0.38	0.69	0.72	0.65
HYENA [Yosef et al., 2012]	0.44	0.49	0.50	0.26	0.43	0.39	0.52	0.54	0.56
DeepWalk [Perozzi et al., 2014]	0.49	0.54	0.53	0.21	0.42	0.39	0.58	0.59	0.61
WSABIE [Yogatama et al., 2015]	0.53	0.57	0.58	0.35	0.55	0.50	0.64	0.66	0.65
PLE [Ren et al., 2016]	0.56	0.60	0.61	0.37	0.57	0.53	0.70	0.71	0.72
CoType	**0.60**	**0.65**	0.66	**0.39**	**0.61**	**0.57**	0.74	**0.76**	**0.75**

2. Performance on Relation Classification. To test the effectiveness of the learned embeddings in representing type semantics of relation mentions, we compare with other methods on classifying the ground-truth relation mention in the evaluation set by target types $\mathbb{R}$. Table 7.7 summarizes the classification accuracy. CoType achieves superior accuracy compared to all other methods and variants (e.g., obtains over 10% enhancement on both the NYT and BioInfer datasets over the next best method). All compared methods (except for MultiR) simply treat $\mathcal{D}_L$ as "perfectly labeled" when training models. The improvement of CoType-RM validates the importance on careful modeling of label noise (i.e., Hypothesis 7.2). Comparing CoType-RM with MultiR, superior performance of CoType-RM demonstrates the effectiveness of partial-label loss over multi-instance learning. Finally, CoType outperforms CoType-RM and CoType-TwoStep validates that the propose translation-based embedding objective is effective in capturing entity-relation cross-constraints.

Table 7.7: Performance comparison on relation classification accuracy over ground-truth relation mentions on the three datasets

Method	NYT	Wiki-KBP	BioInfer
DS+Perceptron [Ling and Weld, 2012]	0.641	0.543	0.470
DS+Kernel [Mooney and Bunescu, 2005]	0.632	0.535	0.419
DeepWalk [Perozzi et al., 2014]	0.580	0.613	0.408
LINE [Tang et al., 2015]	0.765	0.617	0.557
DS+Logistic [Mintz et al., 2009]	0.771	0.646	0.543
MultiR [Hoffmann et al., 2011]	0. 693	0.633	0.501
FCM [Gormley et al., 2015]	0.688	0.617	0.467
CoType-RM	0.812	0.634	0.587
CoType-TwoStep	0.829	0.645	0.591
CoType	**0.851**	**0.669**	**0.617**

3. Performance on Relation Extraction. To test the domain independence of CoType framework, we conduct evaluations on the end-to-end relation extraction. As only MultiR and DS-Joint are able to detection entity and relation mentions in their own framework, we apply our detection results to other compared methods. Table 7.8 shows the evaluation results as well as runtime of different methods. In particular, results at each method's highest F1 score point are reported, after tuning the threshold for each method for determining whether a test mention is None or some target type. Overall, CoType outperforms all other methods on F1 score on all three datasets. We observe that DS-Joint and MultiR suffer from low recall, since their entity detection modules do not work well on $\mathcal{D}_L$ (where many tokens have false negative tags). This demonstrates the effectiveness of the proposed domain-agnostic text segmentation algorithm (see Section 7.3.1). We found that the incremental diagram of learning embedding (i.e., CoType-TwoStep) brings only marginal improvement. In contrast, CoType adopts a "joint mod-

"eling" diagram following Hypothesis 7.3 and achieves significant improvement. In Fig. 7.5, precision-recall curves on NYT and BioInfer datasets further show that CoType can still achieve descent precision with good recall preserved.

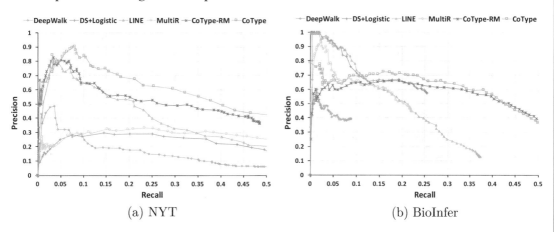

(a) NYT   (b) BioInfer

Figure 7.5: Precision-recall curves of relation extraction on NYT and BioInfer datasets. Similar trend is also observed on the Wiki-KBP dataset.

4. Scalability. In addition to the runtime shown in Table 7.8, Fig. 7.6a tests the scalability of CoType compared with other methods, by running on BioInfer corpora sampled using different ratios. CoType demonstrates a linear runtime trend (which validates our time complexity in Section 7.3.3), and is the only method that is capable of processing the full-size dataset without significant time cost.

7.5 DISCUSSION

1. Example output on news articles. Table 7.9 shows the output of CoType, MultiR, and Logistic on two news sentences from the Wiki-KBP dataset. CoType extracts more relation mentions (e.g., `children`), and predict entity/relation types with better accuracy. Also, CoType can jointly extract typed entity and relation mentions while other methods cannot (or need to do it incrementally).

2. Testing the effect of training corpus size. Figure 7.6b shows the performance trend on Bioinfer dataset when varying the sampling ratio (subset of relation mentions randomly sampled from the training set). F1 scores of all three methods improves as the sampling ratio increases. Co-Type performs best in all cases, which demonstrates its robust performance across corpora of various size.

3. Study the effect of entity type error in relation classification. To investigate the "error propagation" issue of incremental pipeline, we test the changes of relation classification performance

Table 7.8: Performance comparison on end-to-end relation extraction (at the highest F1 point) on the three datasets

Method	NYT [Riedel et al., 2010, Hoffmann et al., 2011]				Wiki-KBP [Ling and Weld, 2012, Ellis et al., 2014]				BioInfer [Pyysalo et al., 2007]			
	Prec	Rec	F1	Time	Prec	Rec	F1	Time	Prec	Rec	F1	Time
DS+Perceptron [Ling and Weld, 2012]	0.068	**0.641**	0.123	15 min	0.233	0.457	0.308	7.7 min	0.357	0.279	0.313	3.3 min
DS+Kernel [Mooney and Bunescu, 2005]	0.095	0.490	0.158	56 hr	0.108	0.239	0.149	9.8 hr	0.333	0.011	0.021	4.2 hr
DS+Logistic [Mintz et al., 2009]	0.258	0.393	0.311	25 min	0.296	0.387	0.335	14 min	**0.572**	0.255	0.353	7.4 min
DeepWalk [Perozzi et al., 2014]	0.176	0.224	0.197	1.1 hr	0.101	0.296	0.150	27 min	0.370	0.058	0.101	8.4 min
LINE [Tang et al., 2015]	0.335	0.329	0.332	2.3 min	0.360	0.257	0.299	1.5 min	0.360	0.275	0.312	35 sec
MultiR [Hoffmann et al., 2011]	0.338	0.327	0.333	5.8 min	0.325	0.278	0.301	4.1 min	0.459	0.221	0.298	2.4 min
FCM [Gormley et al., 2015]	0.553	0.154	0.240	1.3 hr	0.151	**0.500**	0.301	25 min	0.535	0.168	0.255	9.7 min
DS-Joint [Li and Ji, 2014]	**0.574**	0.256	0.354	22 hr	**0.444**	0.043	0.078	54 hr	0.102	0.001	0.002	3.4 hr
CoType-RM	0.467	0.380	0.419	2.6 min	0.342	0.339	0.340	1.5 min	0.482	0.406	0.440	42 sec
CoType-TwoStep	0.368	0.446	0.404	9.6 min	0.347	0.351	0.349	6.1 min	0.502	0.405	0.448	3.1 min
CoType	0.423	0.511	**0.463**	4.1 min	0.348	0.406	**0.369**	2.5 min	0.536	**0.424**	**0.474**	78 sec

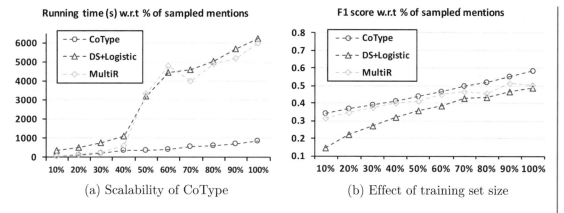

(a) Scalability of CoType   (b) Effect of training set size

Figure 7.6: (a) Scalability study on CoType and the compared methods; and (b) performance changes of relation extraction with respect to sampling ratio of relation mentions on the **Bioinfer** dataset.

by (1) training models without entity types as features; (2) using entity types predicted by FIGER [Ling and Weld, 2012] as features; and (3) using ground-truth ("perfect") entity types as features. Figure 7.7 summarize the accuracy of CoType, its variants and the compared methods. We observe only marginal improvement when using FIGER-predicted types but significant improvement when using ground-truth entity types—this validates the error propagation issue. Moreover, we find that CoType achieves an accuracy close to that of the next best method (i.e., DS + Logistic + Gold entity type). This demonstrates the effectiveness of our proposed joint entity and relation embedding.

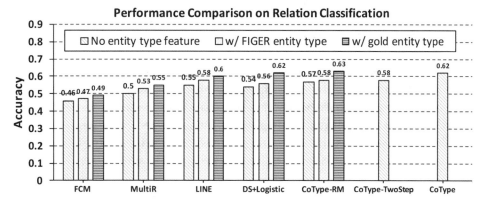

Figure 7.7: Study of entity type error propagation on the **BioInfer** dataset.

Table 7.9: Example output of CoTyᴘᴇ and the compared methods on two news sentences from the **Wiki-KBP** dataset

Text	*Blake Edwards*, a prolific <u>filmmaker</u> who kept alive the tradition of slapstick <u>comedy</u>, <u>died</u> Wednesday of pneumonia at a hospital in *Santa Monica*.	*Anderson* is survived by his wife Carol, <u>sons</u> *Lee* and Albert, daughter Shirley Englebrecht and nine grandchildren.
MultR [Hoffmann et al., 2011]	r^*: person:country_of_birth, $\mathcal{Y}_1^*$: {N/A}, $\mathcal{Y}_2^*$: {N/A}	r^*: None, $\mathcal{Y}_1^*$: {N/A}, $\mathcal{Y}_2^*$: {N/A}
Logistic [Mintz et al., 2009]	r^*: per:country_of_birth, $\mathcal{Y}_1^*$: {person}, $\mathcal{Y}_2^*$: {country}	r^*: None, $\mathcal{Y}_1^*$: {person}, $\mathcal{Y}_2^*$: {person, politician}
CoType	r^*: per:place_of_death, $\mathcal{Y}_1^*$: {person,artist,director}, $\mathcal{Y}_2^*$: {location,city}	r^*: person:children, $\mathcal{Y}_1^*$: {person}, $\mathcal{Y}_2^*$: {person}

7.6 SUMMARY

This work studies domain-independent, joint extraction of typed entities and relations in text with distant supervision. The proposed CoTyᴘᴇ framework runs domain-agnostic segmentation algorithm to mine entity mentions, and formulates the joint entity and relation mention typing problem as a global embedding problem. We design a noise-robust objective to faithfully model noisy type label from distant supervision, and capture the mutual dependencies between entity and relation based on the translation embedding assumption. Experiment results demonstrate the effectiveness and robustness of CoTyᴘᴇ on text corpora of different domains.

Interesting future work includes incorporating pseudo feedback idea [Xu et al., 2013] to reduce false negative type labels in the training data, modeling type correlation in the given type hierarchy [Ren et al., 2016c], and performing type inference for test entity mention and relation mentions jointly. CoTyᴘᴇ relies on minimal linguistic assumption (i.e., only POS-tagged corpus is required) and thus can be extended to different languages where pre-trained POS taggers is available.

C H A P T E R   8

Pattern-Enhanced Embedding Learning for Relation Extraction

Meng Qu, *Department of Computer Science, University of Illinois at Urbana-Champaign*

Relation extraction is an important task in data mining and natural language processing. Given a text corpus, relation extraction aims at extracting a set of relation instances (i.e., a pair of entities and their relation) based on some given examples. Many efforts [Culotta and Sorensen, 2004, Mooney and Bunescu, 2006, Ren et al., 2017a] have been done on sentence-level relation extraction, where the goal is to predict the relation for a pair of entities mentioned in a sentence (e.g., predict the relation between "*Beijing*" and "*China*" in sentence 1 of Fig. 8.1).

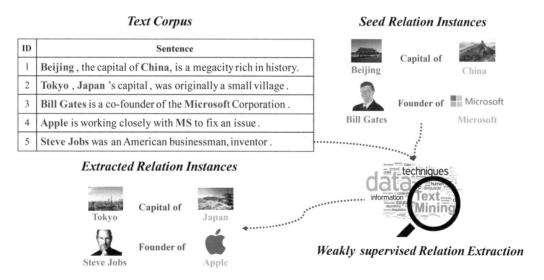

Figure 8.1: Illustration of weakly supervised relation extraction. Given a text corpus and a few relation instances as seeds, the goal is to extract more instances from the corpus.

8.1   OVERVIEW AND MOTIVATION

Despite its wide applications, these studies usually require a large number of human-annotated sentences as training data, which are expensive to obtain. In many cases (e.g., KB completion [Xu et al., 2014]), it is also desirable to extract a set of relation instances by consolidating evidences from multiple sentences in corpora, which cannot be directly achieved by these studies. Instead of looking at individual sentences, corpus-level relation extraction [Bing et al., 2015, Hoffmann et al., 2011, Mintz et al., 2009, Riedel et al., 2013, Zeng et al., 2015] identifies relation instances from text corpora using evidences from multiple sentences. This also makes it possible to apply weakly supervised methods based on corpus-level statistics [Agichtein and Gravano, 2000, Curran et al., 2007]. Such weakly supervised approaches usually take a few relation instances as seeds, and extract more instances by consolidating redundant information collected from large corpora. The extracted instances can serve as extra knowledge in various downstream applications, including KB completion [Riedel et al., 2013, Toutanova et al., 2015], corpus-level relation extraction [Lin et al., 2016, Zeng et al., 2015], hypernym discovery [Shwartz et al., 2016, Snow et al., 2005], and synonym discovery [Qu et al., 2017, Wang et al., 2016].

8.1.1   CHALLENGES

In this work, we focus on corpus-level relation extraction in the *weakly supervised setting*. There are broadly two types of weakly supervised approaches for corpus-level relation extraction. Among them, pattern-based approaches predict the relation of an entity pair from multiple sentences mentioning both entities. To do that, traditional approaches [Nakashole et al., 2012, Schmitz et al., 2012, Yahya et al., 2014] extract textual patterns (e.g., tokens between a pair of entities) and new relation instances in a bootstrapping manner. However, many relations could be expressed in a variety of ways. Due to such diversity, these approaches often have difficulty matching the learned patterns to unseen contexts, leading to the problem of semantic drift [Curran et al., 2007] and inferior performance. For example, with the given instance *"(Beijing, Capital of, China)"* in Fig. 8.1, *"[Head], the capital of [Tail]"* will be extracted as a textual pattern from sentence 1. But we have difficulty in matching the pattern to sentence 2 even though both sentences refer to the same relation *"Capital of."* Recent approaches [Liu et al., 2015, Xu et al., 2015] try to overcome the sparsity issue of textual patterns by encoding textual patterns with neural networks, so that pattern matching can be replaced by similarity measurement between vector representations. However, these approaches typically rely on large amount of labeled instances to train effective models [Shwartz et al., 2016], making it hard to deal with the weakly supervised setting.

Alternatively, distributional approaches resort to the corpus-level co-occurrence statistics of entities. The basic idea is to learn low-dimensional representations of entities to preserve such statistics, so that entities with similar semantic meanings tend to have similar representations. With entity representations, a relation classifier can be learned using the labeled relation instances, which takes entity representations as features and predicts the relation of a pair of entities. To learn entity representations, some approaches [Mikolov et al., 2013, Pennington et al., 2014, Tang et al., 2015] only consider the given text corpus. Despite the unsupervised property, their performance is usually limited due to the lack of supervision [Xu et al., 2014]. To learn more effective representations for relation extraction, some other approaches [Wang et al., 2014, Xu et al., 2014] jointly learn entity representations and relation classification using the labeled instances. However, similar to pattern-based approaches, distributional approaches also require considerable amount of relation instances to achieve good performance [Xu et al., 2014], which are usually hard to obtain in the weakly supervised setting.

8.1.2   PROPOSED SOLUTION

The pattern and distributional-based approaches extract relations from different perspectives, which are naturally complementary to each other. Ideally, we would wish to integrate both approaches, so that they can mutually enhance and reduce the reliance on the given relation instances. Toward integrating both approaches, several existing studies [Qu et al., 2017, Shwartz et al., 2016, Toutanova et al., 2015] try to jointly train a distributional model and a pattern model using the labeled instances. However, the supervision of their frameworks still totally comes from the given relation instances, which is insufficient in the weakly supervised setting. Therefore, their performance is yet far from satisfaction, and we are seeking an approach that is more robust to the scarcity of seed instances.

In this chapter, we propose such an approach called REPEL (Relation Extraction with Pattern-enhanced Embedding Learning) to weakly supervised relation extraction. Our approach consists of a pattern module and a distributional module (see Fig. 8.2). The pattern module aims at learning a set of reliable textual patterns for relation extraction, while the distributional module tries to learn a relation classifier on entity representations for prediction. Different from existing studies, we follow the co-training [Blum and Mitchell, 1998] strategy and encourage both modules to provide extra supervision for each other, which is expected to complement the limited supervision from the given seed instances (see Fig. 8.3). Specifically, the pattern module acts as a *generator*, as it can extract some candidate instances based on the discovered reliable patterns, whereas the distributional module is treated as a *discriminator* to evaluate the quality of each generated instance, that is, whether an instance is reasonable. To encourage the collaboration of both modules, we formulate a joint optimization process, in which we iterate between two sub-processes. In the first sub-process, the discriminator (distributional module) will evaluate the instances generated by the generator (pattern module), and the results serve as extra signals to adjust the generator. In the second sub-process, the generator (pattern module) will in turn generate a set of highly confident instances, which serve as extra training seeds to improve the discriminator (distributional module). During training, we keep iterating between the two sub-processes, so that both modules can be consistently improved. Once the training converges, both modules can be applied to relation extraction, which extract new relation instances from different perspectives.

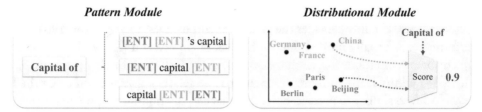

Figure 8.2: Illustration of the modules. The pattern module aims to learn reliable textual patterns for each relation. The distributional module tries to learn entity representations and a score function to estimate the quality of each instance.

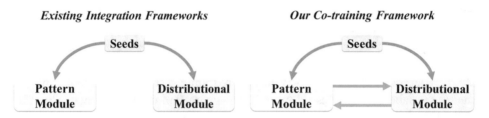

Figure 8.3: Comparison with existing integration frameworks. Existing frameworks totally rely on the seed instances to provide supervision. Our framework encourages both modules to provide extra supervision for each other.

8.2 THE REPEL FRAMEWORK

In this chapter, we propose a co-training based framework called REPEL. Our framework consists of two modules, a pattern module and a distributional module (see Fig. 8.2), which extract relations from different perspectives. The pattern module aims at finding a set of reliable textual patterns for relation extraction, while the distributional module tries to learn entity representations and train a score function, which serves as a relation classifier to measure the quality of a relation instance. Different from existing studies, both modules are encouraged to provide extra supervision to each other, which is expected to complement the limited supervision from seed instances (see Fig. 8.3). Specifically, the pattern module is treated as a generator since it can generate some candidate relation instances, and meanwhile the distributional module acts as a discriminator to evaluate each instance. During training, the discriminator evaluates the instances generated by the generator, and the results serve as extra signals to adjust the generator. On the other hand, the generator will in turn generate some highly confident instances, which act as extra seeds to improve the discriminator. We keep iterating between adjusting the pattern module and improving the distributional module. Once the training process converges, both modules can be adopted to discover more instances.

The overall objective function is summarized below:

$$O = \max_{P,D} \left\{ O_p + O_d + \lambda O_i \right\}. \qquad (8.1)$$

In the objective function, P represents the parameters of the pattern module, that is, a given number of reliable patterns for each target relation. D denotes the parameters of the distributional module, that is, entity representations and a score function that serves as a relation classifier. The objective function consists of three terms. Among them, O_p is the objective of the pattern module, in which we leverage the given seed instances for pattern selection. O_d is the objective of the distributional module, which learns relevant parameters under the guidance of seed instances. Finally, O_i models the interactions of both modules.

8.3   EXPERIMENT

In this section, we evaluate our approach on two downstream applications: KB completion with text corpora (KBC) and corpus-level relation extraction (RE). In KB completion with text corpora, the key task is to predict the missing relationships between each pair of entities in KBs. Since some pairs of entities may not co-occur in any sentences of the given corpus, the learned pattern module cannot provide information for predicting their relations. Therefore, for KBC we only use the entity representations and score function learned by the distributional module for extraction, and we expect to show that the pattern module can provide extra seeds during training, yielding a more effective distributional module. For corpus-level RE, it aims at predicting the relation of a pair of entities from several sentences mentioning both of them. In this case, the reliable patterns learned by the pattern module can capture the local context information from the sentences. Therefore, we focus on utilizing the learned pattern module for prediction in RE, and we expect to show that the distributional module can enhance the pattern module by providing extra supervision to select reliable patterns.

1. Knowledge Base Completion with Text Corpora (KBC). We present the quantitative results in Table 8.1. For the approach only considering the given seed instances (TransE), we see the performance is very limited due to the scarcity of seeds. Along the other line, the approach considering text corpora (word2vec) achieves relatively better results, but are still far from satisfactory, since it ignores the supervision from the seed instances. If we consider both the text corpus and seed instances for entity representation learning (RK), we obtain much better results. Moreover, by further jointly training a pattern model (DPE, CONV), the hits ratio can be further significantly improved.

For our proposed approach, with only the distributional module (REPEL-D), it already outperforms all the baseline approaches. Compared with DPE, the performance gain of REPEL-D mainly comes from the usage of the score function which utilizing seed instances, which can better model different relations. Compared with CONV, REPEL-D achieves better results, as the distributional information in text corpora can be better captured. Moreover, by

Table 8.1: Quantitative results on the KBC task

Algorithm	Wiki + Freebase		NYT + Freebase	
	Hits@10	MR	Hits@10	MR
TransE [Bordes et al., 2013]	7.13	4328.40	15.94	3833.47
word2vec [Mikolov et al., 2013]	32.12	203.53	15.56	913.04
RK [Wang et al., 2016]	41.49	72.87	29.01	307.89
DPE [Qu et al., 2017]	45.45	78.87	32.47	279.99
CONV [Toutanova et al., 2015]	46.84	139.81	31.51	903.38
REPEL-D	47.49	67.28	35.79	234.23
REPEL	**51.18**	**62.18**	**38.98**	**199.44**

encouraging the collaboration of both modules (REPEL), the results are further significantly improved. This observation demonstrates that the pattern module can indeed help improve the distributional module by providing some highly confident instances.

Overall, our approach achieves quite impressive results on the KB completion task compared with several strong baseline approaches. Also, the pattern module can indeed enhance the distributional module with our co-training framework.

2. Corpus-level Relation Extraction (RE). Next, we show the results on the corpus-level relation extraction task. We present the quantitative results in Table 8.2. For the approaches using textual patterns (PATTY, Snowball), we see the results are quite limited especially on the NYT dataset. This is because it discovers informative patterns in a bootstrapping way, which can lead to the semantic drift problem [Curran et al., 2007] and thus harm the performance. For other neural network-based pattern approaches (PathCNN, CNN-ATT, PCNN-ATT), although they are proved to be very effective when the given instances are abundant, their performance in the weakly supervised setting is not satisfactory. The reason is that they typically deploy complicated convolutional layers or recurrent layers in their model, which rely on massive relation instances to tune. However, in our setting, the instances are very limited, leading to their poor performance. For the integration approach (LexNET), although it incorporates the distributional information, the performance is still quite limited especially on the NYT dataset. This is because the joint training framework of LexNET also requires considerable training instances.

Overall, in the weakly supervised setting, our approach is able to achieve comparable results compared with the neural methods. Besides, the distributional module can indeed improve the pattern module with our co-training framework.

Table 8.2: Quantitative results on the RE task

Algorithm	Wiki + Freebase						NYT + Freebase					
	P@50	R@50	F1@50	P@100	R@100	F1@100	P@50	R@50	F1@50	P@100	R@100	F1@100
Snowball [Agichtein and Gravano, 2000]	58.00	22.14	32.05	65.00	49.62	56.28	20.00	4.50	7.35	21.00	9.46	13.04
PATTY [Nakashole et al., 2012]	60.00	22.90	33.15	61.00	46.56	52.81	28.00	6.31	10.30	20.00	9.01	12.42
CNN-ATT [Lin et al., 2016]	26.00	9.92	14.36	22.00	16.79	19.05	24.00	5.41	8.83	29.00	13.06	18.01
PCNN-ATT [Lin et al., 2016]	58.00	22.14	32.05	36.00	27.48	31.17	46.00	10.36	16.91	26.00	11.71	16.15
PathCNN [Zeng et al., 2017]	36.00	13.74	19.89	38.00	29.01	32.90	42.00	9.46	15.44	26.00	11.71	16.15
LexNET [Schwartz et al., 2016; Schwartz and Dagan, 2016]	74.00	28.24	40.88	61.00	46.56	52.81	32.00	7.21	11.77	26.00	11.71	16.15
REPEL-D	14.00	5.34	7.73	17.00	12.98	14.72	6.00	1.35	2.20	7.00	3.15	4.34
REPEL-P	64.00	24.43	35.36	70.00	53.44	60.61	32.00	7.21	11.77	33.00	14.86	20.49
REPEL	78.00	29.77	43.09	76.00	58.02	65.80	48.00	10.81	17.65	43.00	19.37	26.71

8.4 SUMMARY

In this chapter, we introduce corpus-level relation extraction in the weakly supervised setting. We proposed a novel co-training framework called REPEL to integrate a pattern module and a distributional module. Our framework encouraged both modules to provide extra supervision for each other, so that they can collaborate to overcome the scarcity of seeds. Experimental results proved the effectiveness of our framework. In the future, we plan to enhance the pattern module by using neural models for pattern encoding.

CHAPTER 9

Heterogeneous Supervision for Relation Extraction

Liyuan Liu, *Department of Computer Science, University of Illinois at Urbana-Champaign*

One of the most important tasks toward text understanding is to detect and categorize semantic relations between two entities in a given context. Typically, existing methods follow the supervised learning paradigm, and require extensive annotations from domain experts, which are costly and time-consuming. To alleviate such drawback, attempts have been made to build relation extractors with a small set of seed instances or human-crafted patterns [Carlson et al., 2010, Nakashole et al., 2011], based on which more patterns and instances will be iteratively generated by bootstrap learning. However, these methods often suffer from semantic drift [Mintz et al., 2009]. Besides, KBs like Freebase have been leveraged to automatically generate training data and provide distant supervision [Mintz et al., 2009]. Nevertheless, for many domain-specific applications, distant supervision is either non-existent or insufficient (usually less than 25% of relation mentions are covered [Ling and Weld, 2012, Ren et al., 2015]).

9.1 OVERVIEW AND MOTIVATION

Only recently have preliminary studies been developed to unite different supervisions, including KBs and domain specific patterns, which are referred to as *heterogeneous supervision*. As shown in Fig 9.1, these supervisions often conflict with each other [Ratner et al., 2016]. To address these conflicts, data programming [Ratner et al., 2016] employs a generative model, which encodes supervisions as *labeling functions*, and adopts the source consistency assumption: *a source is likely to provide true information with the same probability for all instances*. This assumption is widely used in true label discovery literature [Li et al., 2016] to model reliabilities of information sources like crowdsourcing and infer the true label from noisy labels. Accordingly, most true label discovery methods would trust a human annotator on all instances to the same level.

However, labeling functions, unlike human annotators, do not make casual mistakes but follow certain "error routine." Thus, the reliability of a labeling function is not consistent among different pieces of instances. In particular, a labeling function could be more reliable for a certain subset [Varma et al., 2016] (also known as its *proficient subset*) comparing to the rest. We identify

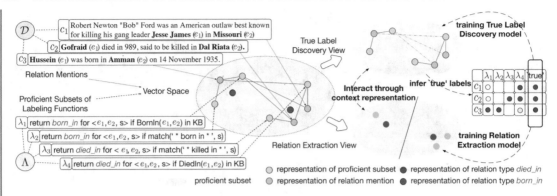

Figure 9.1: REHession Framework.

these proficient subsets based on context information, only trust labeling functions on these subsets and avoid assuming global source consistency.

Meanwhile, embedding methods have demonstrated great potential in capturing semantic meanings, which also reduce the dimension of overwhelming text features. Here, we present REHession, a novel framework capturing context's semantic meaning through representation learning, and conduct both relation extraction and true label discovery in a context-aware manner. Specifically, as depicted in Fig. 9.1, we embed relation mentions in a low-dimension vector space, where similar relation mentions tend to have similar relation types and annotations. "True" labels are further inferred based on the reliability of labeling functions, which are calculated with their proficient subsets' representations. Then, these inferred true labels would serve as supervision for all components, including context representation, true label discovery, and relation extraction. Besides, the context representation bridges relation extraction with true label discovery, and allows them to enhance each other.

9.2 PRELIMINARIES

In this section, we would formally define relation extraction and heterogeneous supervision, including the format of labeling functions.

9.2.1 RELATION EXTRACTION

Here we conduct relation extraction in *sentence-level* [Bao et al., 2014]. For a sentence d, an entity mention is a token span in d which represents an entity, and a relation mention is a triple (e_1, e_2, d) which consists of an ordered entity pair (e_1, e_2) and d. And the relation extraction task is to categorize relation mentions into a given set of relation types $\mathbb{R}$, or Not-Target-Type (None) which means the type of the relation mention does not belong to $\mathbb{R}$.

9.2.2   HETEROGENEOUS SUPERVISION

Similar to Ratner et al. [2016], we employ labeling functions as basic units to encode supervision information and generate annotations. Since different supervision information may have different proficient subsets, we require each labeling function to encode only one elementary supervision information. Specifically, in the relation extraction scenario, we require each labeling function to only annotate one relation type based on one elementary piece of information, e.g., four examples are listed in Fig. 9.1.

Notice that knowledge-based labeling functions are also considered to be noisy because relation extraction is conducted in sentence-level, e.g., although `president_of` (*Obama*, *USA*) exists in KB, it should not be assigned with "*Obama* was born in Honolulu, Hawaii, *U.S.*," since `president_of` is irrelevant to the context.

9.2.3   PROBLEM DEFINITION

For a POS-tagged corpus $\mathcal{D}$ with detected entities, we refer its relation mentions as $\mathcal{C} = \{c_i = (e_{i,1}, e_{i,2}, d), \forall d \in \mathcal{D}\}$. Our goal is to annotate entity mentions with relation types of interest ($\mathbb{R} = \{r_1, \ldots, r_K\}$) or `None`. We require users to provide heterogeneous supervision in the form of labeling function $\Lambda = \{\lambda_1, \ldots, \lambda_M\}$, and mark the annotations generated by Λ as $\mathcal{O} = \{o_{c,i}|\lambda_i$ generate annotation $o_{c,i}$ for $c \in \mathcal{C}\}$. We record relation mentions annotated by Λ as $\mathcal{C}_l$, and refer relation mentions without annotation as $\mathcal{C}_u$. Then, our task is to train a relation extractor based on $\mathcal{C}_l$ and categorize relation mentions in $\mathcal{C}_u$.

9.3   THE REHESSION FRAMEWORK

Here, we present REHESSION, a novel framework to infer true labels from automatically generated noisy labels, and categorize unlabeled instances into a set of relation types. Intuitively, errors of annotations ($\mathcal{O}$) come from mismatch of contexts (e.g., in Fig. 9.1, λ_1 annotates c_1 and c_2 with "true" labels but for mismatched contexts "killing" and "killed"). Accordingly, we should only trust labeling functions on *matched* context, e.g., trust λ_1 on c_3 due to its context "was born in," but not on c_1 and c_2. On the other hand, relation extraction can be viewed as *matching* appropriate relation type to a certain context. These two *matching* processes are closely related and can enhance each other, while context representation plays an important role in both of them.

Framework Overview. We propose a general framework to learn the relation extractor from automatically generated noisy labels. As plotted in Fig. 9.1, distributed representation of context bridges relation extraction with true label discovery, and allows them to enhance each other. Specifically, it follows the steps below.

1. After being extracted from context, text features are embedded in a low dimension space by representation learning (see Fig. 9.2).

2. Text feature embeddings are utilized to calculate relation mention embeddings (see Fig. 9.2).

3. With relation mention embeddings, true labels are inferred by calculating labeling functions' reliabilities in a context-aware manner (see Fig. 9.1).

4. Inferred true labels would "supervise" all components to learn model parameters (see Fig. 9.1).

We now proceed by introducing these components of the model in further details.

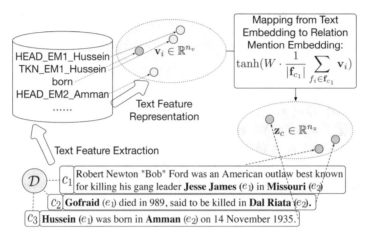

Figure 9.2: Relation mention representation.

9.3.1   MODELING RELATION MENTION

As shown in Table 9.1, we extract abundant lexical features [Mintz et al., 2009, Ren et al., 2016e] to characterize relation mentions. However, this abundance also results in the gigantic dimension of original text features ($\sim 10^7$ in our case). In order to achieve better generalization ability, we represent relation mentions with low dimensional ($\sim 10^2$) vectors. In Fig. 9.2, for example, relation mention c_3 is first represented as bag-of-features. After learning text feature embeddings, we use the average of feature embedding vectors to derive the embedding vector for c_3.

Text Feature Representation. Similar to other principles of embedding learning, we assume text features occurring in the same contexts tend to have similar meanings (also known as distributional hypothesis [Harris, 1954]). Furthermore, we let each text feature's embedding vector to predict other text features occurred in the same relation mentions or context. Thus, text features with similar meaning should have similar embedding vectors. Formally, we mark text features as $\mathcal{F} = \{f_1, \cdots, f_{|\mathcal{F}|}\}$, record the feature set for $\forall c \in \mathcal{C}$ as $_c$, and represent the embedding vector for

Table 9.1: Text features $\mathcal{F}$ used in this chapter. (*"Hussein," "Amman," "**Hussein** was born in **Amman**"*) is used as an example.

Feature	Description	Example
Entity mention (EM) head	Syntactic head token of each entity mention	*"HEAD_EM1_Hussein"*, …
Entity Mention Token	Tokens in each entity mention	*"TKN_EM1_Hussein"*, …
Tokens between two EMs	Tokens between two EMs	*"was", "born", "in"*
Part-of-speech (POS) tag	POS tags of tokens between two EMs	*"VBD", "VBN", "IN"*
Collocations	Bigrams in left/right 3-word window of each EM	*"Hussein was", "in Amman"*
Entity mention order	Whether EM 1 is before EM 2	*"EM1_BEFORE_EM2"*
Entity mention distance	Number of tokens between the two EMs	*"EM_DISTANCE_3"*
Body entity mentions numbers	Number of EMs between the two EMs	*"EM_NUMBER_0"*
Entity mention context	Unigrams before and after each EM	*"EM_AFTER_was"*, …
Brown cluster (learned on $\mathcal{D}$)	Brown cluster ID for each token	*"Brown_010011001"*, …

f_i as $\mathbf{v}_i \in \mathbb{R}^{n_v}$, and we aim to maximize the following log likelihood: $\sum_{c \in \mathcal{C}_l} \sum_{f_i, f_j \in c} \log p(f_i | f_j)$, where $p(f_i | f_j) = \exp(\mathbf{v}_i^T \mathbf{v}_j^*) / \sum_{f_k \in \mathcal{F}} \exp(\mathbf{v}_i^T \mathbf{v}_k^*)$.

However, the optimization of this likelihood is impractical because the calculation of $\nabla p(f_i | f_j)$ requires summation over all text features, whose size exceeds 10^7 in our case. In order to perform efficient optimization, we adopt the negative sampling technique [Mikolov et al., 2013] to avoid this summation. Accordingly, we replace the log likelihood with Eq. (9.1) as below:

$$\mathcal{J}_E = \sum_{\substack{c \in \mathcal{C}_l \\ f_i, f_j \in c}} \left(\log \sigma \left(\mathbf{v}_i^T \mathbf{v}_j^* \right) - \sum_{k=1}^{V} \mathcal{E}_{f_{k'} \sim \hat{P}} \left[\log \sigma \left(-\mathbf{v}_i^T \mathbf{v}_{k'}^* \right) \right] \right), \qquad (9.1)$$

where $\hat{P}$ is noise distribution used in Mikolov et al. [2013], σ is the sigmoid function, and V is number of negative samples.

Relation Mention Representation. With text feature embeddings learned by Eq. (9.1), a naive method to represent relation mentions is to concatenate or average its text feature embeddings. However, text features embedding may be in a different semantic space with relation types. Thus, we directly learn a mapping g from text feature representations to relation mention representations [Gysel et al., 2016a,b] instead of simple heuristic rules like concatenate or average (see Fig. 9.2):

$$tz_c = g(c) = \tanh\left(\mathbf{W} \cdot \frac{1}{|c|} \sum_{f_i \in c} v_i\right),\qquad(9.2)$$

where tz_c is the representation of $c \in C_l$, W is a $n_z \times n_v$ matrix, n_z is the dimension of relation mention embeddings, and tanh is the element-wise hyperbolic tangent function.

In other words, we represent bag of text features with their average embedding, then apply linear map and hyperbolic tangent to transform the embedding from text feature semantic space to relation mention semantic space. The nonlinear tanh function allows nonlinear class boundaries in other components, and also regularize relation mention representation to range $[-1, 1]$ which avoids numerical instability issues.

9.3.2 TRUE LABEL DISCOVERY

Because heterogeneous supervision generates labels in a discriminative way, we suppose its errors follow certain underlying principles, i.e., if a labeling function annotates a instance correctly/wrongly, it would annotate other similar instances correctly/wrongly. For example, λ_1 in Fig. 9.1 generates wrong annotations for two similar instances c_1, c_2 and would make the same errors on other similar instances. Since context representation captures the semantic meaning of relation mention and would be used to identify relation types, we also use it to identify the mismatch of context and labeling functions. Thus, we suppose for each labeling function λ_i, there exists a proficient subset S_i on $\mathbb{R}^{n_z}$, containing instances that λ_i can precisely annotate. In Fig. 9.1, for instance, c_3 is in the proficient subset of λ_1, while c_1 and c_2 are not. Moreover, the generation of annotations are not really random, and we propose a probabilistic model to describe the level of mismatch from labeling functions to real relation types instead of annotations' generation.

We assume the indicator of whether c belongs to S_i, $s_{c,i} = \delta(c \in S_i)$, would first be generated based on context representation

$$p\left(s_{c,i} = 1 | tz_c, l_i\right) = p\left(c \in S_i\right) = \sigma\left(tz_c^T l_i\right).\qquad(9.3)$$

Then the correctness of annotation $o_{c,i}$, $\rho_{c,i} = \delta(o_{c,i} = o_c^*)$, would be generated. Furthermore, we assume $p(\rho_{c,i} = 1 | s_{c,i} = 1) = \phi_1$ and $p(\rho_{c,i} = 1 | s_{c,i} = 0) = \phi_0$ to be constant for all relation mentions and labeling functions.

Because $s_{c,i}$ would not be used in other components of our framework, we integrate out $s_{c,i}$ and write the log likelihood as

$$\mathcal{J}_T = \sum_{o_{c,i} \in \mathcal{O}} \log\Big(\sigma\left(tz_c^T l_i\right) \phi_1^{\delta(o_{c,i}=o_c^*)} (1-\phi_1)^{\delta(o_{c,i}\neq o_c^*)}$$
$$+ \left(1 - \sigma\left(tz_c^T l_i\right)\right) \phi_0^{\delta(o_{c,i}=o_c^*)} (1-\phi_0)^{\delta(o_{c,i}\neq o_c^*)}\Big).\qquad(9.4)$$

Note that o_c^* is a hidden variable but not a model parameter, and $\mathcal{J}_T$ is the likelihood of $\rho_{c,i} = \delta(o_{c,i} = o_c^*)$. Thus, we would first infer $o_c^* = \mathrm{argmax}_{o_c^*} \mathcal{J}_T$, then train the true label discovery model by maximizing $\mathcal{J}_T$.

9.3.3   MODELING RELATION TYPE

We now discuss the model for identifying relation types based on context representation. For each relation mention c, its representation tz_c implies its relation type, and the distribution of relation type can be described by the soft-max function:

$$p\left(r_i | tz_c\right) = \frac{\exp\left(tz_c^T \mathbf{t}_i\right)}{\sum_{r_j \in \mathbb{R} \cup \{\texttt{None}\}} \exp\left(tz_c^T \mathbf{t}_j\right)}, \tag{9.5}$$

where $\mathbf{t}_i \in \mathbb{R}^{v_z}$ is the representation for relation type r_i. Moreover, with the inferred true label o_c^*, the relation extraction model can be trained as a multi-class classifier. Specifically, we use Eq. (9.5) to approach the distribution

$$p\left(r_i | o_c^*\right) = \begin{cases} 1 & r_i = o_c^* \\ 0 & r_i \neq o_c^*. \end{cases} \tag{9.6}$$

Moreover, we use KL-divergence to measure the dissimilarity between two distributions, and formulate model learning as maximizing $\mathcal{J}_R$:

$$\mathcal{J}_R = -\sum_{c \in \mathcal{C}_l} KL\left(p\left(.|tz_c\right) \| p\left(.|o_c^*\right)\right), \tag{9.7}$$

where $KL(p(.|tz_c) \| p(.|o_c^*))$ is the KL-divergence from $p(r_i|o_c^*)$ to $p(r_i|tz_c)$, $p(r_i|tz_c)$ and $p(r_i|o_c^*)$ has the form of Eqs. (9.5) and (9.6).

9.3.4   MODEL LEARNING

Based on Eqs. (9.1), (9.4), and (9.7), we form the joint optimization problem for model parameters as

$$\min_{W, \mathbf{v}, \mathbf{v}^*, \mathbf{l}, \mathbf{t}, o*} \mathcal{J} = -\mathcal{J}_R - \lambda_1 \mathcal{J}_E - \lambda_2 \mathcal{J}_T$$
$$\text{s.t. } \forall c \in \mathcal{C}_l, o_c^* = \mathrm{argmax}_{o_c^*} \mathcal{J}_T, tz_c = g(\mathbf{c}). \tag{9.8}$$

Collectively optimizing Eq. (9.8) allows heterogeneous supervision guiding all three components, while these components would refine the context representation, and enhance each other.

In order to solve the joint optimization problem in Eq. (9.8) efficiently, we adopt the stochastic gradient descent algorithm to update $\{W, \mathbf{v}, \mathbf{v}^*, \mathbf{l}, \mathbf{t}\}$ iteratively, and o_c* is estimated by maximizing $\mathcal{J}_T$ after calculating tz_c. Additionally, we apply the widely used dropout techniques [Srivastava et al., 2014] to prevent overfitting and improve generalization performance.

The learning process of REHession is summarized as below. In each iteration, we would sample a relation mention c from $\mathcal{C}_l$, then sample c's text features and conduct the text features' representation learning. After calculating the representation of c, we would infer its true label o_c^* based on our true label discovery model, and finally update model parameters based on o_c^*.

9.3.5   RELATION TYPE INFERENCE

We now discuss the strategy of performing type inference for $\mathcal{C}_u$. We observe that the proportion of None in $\mathcal{C}_u$ is usually much larger than in $\mathcal{C}_l$. Additionally, not like other relation types in $\mathbb{R}$, None does not have a coherent semantic meaning. Similar to Ren et al. [2016e], we introduce a heuristic rule: identifying a relation mention as None when (1) our relation extractor predict it as None, or (2) the entropy of $p(.|tz_c)$ over $\mathbb{R}$ exceeds a pre-defined threshold η. The entropy is calculated as $H(p(.|tz_c)) = -\sum_{r_i \in \mathbb{R}} p(r_i|tz_c)\log(p(r_i|tz_c))$. And the second situation means based on relation extractor this relation mention is not likely belonging to any relation types in $\mathbb{R}$.

9.4   EXPERIMENTS

In this section, we empirically validate our method by comparing to the state-of-the-art relation extraction methods on news and Wikipedia articles.

Given the experimental setup described above, the averaged evaluation scores in ten runs of relation classification and relation extraction on two datasets are summarized in Table 9.2.

From the comparison, it shows that NL strategy yields better performance than TD strategy, since the true labels inferred by investment are actually wrong for many instances. On the other hand, our method introduces context-awareness to true label discovery, while the inferred true label guides the relation extractor achieving the best performance. This observation justifies the motivation of avoiding the source consistency assumption and the effectiveness of proposed true label discovery model.

One could also observe the difference between REHession and the compared methods is more significant on the NYT dataset than on the Wiki-KBP dataset. This observation accords with the fact that the NYT dataset contains more conflicts than KBP dataset, and the intuition is that our method would have more advantages on more conflicting labels.

Among four tasks, the relation classification of Wiki-KBP dataset has highest label quality, i.e., conflicting label ratio, but with least number of training instances. And CoType-RM and DSL reach relatively better performance among all compared methods. CoType-RM performs much better than DSL on Wiki-KBP relation classification task, while DSL gets better or similar performance with CoType-RM on other tasks. This may be because the representation learning method is able to generalize better, thus performs better when the training set size is small. However, it is rather vulnerable to the noisy labels compared to DSL. Our method employs embedding techniques, and also integrates context-aware true label discovery to de-noise labels, making the embedding method rather robust, thus achieves the best performance on all tasks.

Table 9.2: Performance comparison of relation extraction and relation classification

| Method | Relation Extraction | | | | | | Relation Classification | |
| | NYT | | | Wiki-KBP | | | NYT | Wiki-KBP |
	Prec	Rec	F1	Prec	Rec	F1	Accuracy	Accuracy
NL+FIGER	0.2364	0.2914	0.2606	0.2048	0.4489	0.2810	0.6598	0.6226
NL+BFK	0.1520	0.0508	0.0749	0.1504	0.3543	0.2101	0.6905	0.5000
NL+DSL	0.4150	0.5414	0.4690	0.3301	0.5446	0.4067	0.7954	0.6355
NL+MultiR	0.5196	0.2755	0.3594	0.3012	0.5296	0.3804	0.7059	0.6484
NL+FCM	0.4170	0.2890	0.3414	0.2523	0.5258	0.3410	0.7033	0.5419
NL+CoType-RM	0.3967	0.4049	0.3977	**0.3701**	0.4767	0.4122	0.6485	0.6935
TD+FIGER	0.3664	0.3350	0.3495	0.2650	**0.5666**	0.3582	0.7059	0.6355
TD+BFK	0.1011	0.0504	0.0670	0.1432	0.1935	0.1646	0.6292	0.5032
TD+DSL	0.3704	0.5025	0.4257	0.2950	0.5757	0.3849	0.7570	0.6452
TD+MultiR	**0.5232**	0.2736	0.3586	0.3045	0.5277	0.3810	0.6061	0.6613
TD+FCM	0.3394	0.3325	0.3360	0.1964	0.5645	0.2914	0.6803	0.5645
TD+CoType-RM	0.4516	0.3499	0.3923	0.3107	0.5368	0.3879	0.6409	0.6890
REHession	0.4122	**0.5726**	**0.4792**	0.3677	0.4933	**0.4208**	**0.8381**	**0.7277**

9.5   SUMMARY

In this chapter, we propose REHESSION, an embedding framework to extract relation under heterogeneous supervision. When dealing with heterogeneous supervisions, one unique challenge is how to resolve conflicts generated by different labeling functions. Accordingly, we go beyond the "source consistency assumption" in prior works and leverage context-aware embeddings to induce proficient subsets. The resulting framework bridges true label discovery and relation extraction with context representation, and allows them to mutually enhance each other. Experimental evaluation justifies the necessity of involving context-awareness, the quality of inferred true label, and the effectiveness of the proposed framework on two real-world datasets.

CHAPTER 10

Indirect Supervision: Leveraging Knowledge from Auxiliary Tasks

Zeqiu Wu, *Department of Computer Science, University of Illinois at Urbana-Champaign*

Typically, relation extraction (RE) systems rely on training data, primarily acquired via human annotation, to achieve satisfactory performance. However, such manual labeling process can be costly and non-scalable when adapting to other domains (e.g., biomedical domain). In addition, when the number of types of interest becomes large, the generation of handcrafted training data can be error-prone. To alleviate such an exhaustive process, the recent trend has deviated toward the adoption of distant supervision (DS).

10.1 OVERVIEW AND MOTIVATION

DS replaces the manual training data generation with a pipeline that automatically links texts to a KB. The pipeline has the following steps: (1) detect entity mentions in text; (2) map detected entity mentions to entities in KB; and (3) assign, to the candidate type set of each entity mention pair, all KB relation types between their KB-mapped entities. However, the noise introduced to the automatically generated training data is not negligible. There are two major causes of error: incomplete KB and context-agnostic labeling process. If we treat unlinkable entity pairs as the pool of negative examples, false negatives can be commonly encountered as a result of the insufficiency of facts in KBs, where many true entity or relation mentions fail to be linked to KBs (see example in Fig. 10.1). In this way, models counting on extensive negative instances may suffer from such misleading training data. On the other hand, context-agnostic labeling can engender false positive examples, due to the inaccuracy of the DS assumption that if a sentence contains any two entities holding a relation in the KB, the sentence must be expressing such relation between them. For example, entities *"Donald Trump"* and *"United States"* in the sentence *"Donald Trump flew back to United States"* can be labeled as `president_of` as well as "born_in," although only an out-of-interest relation type "travel_to" is expressed explicitly (as shown in Fig. 10.1).

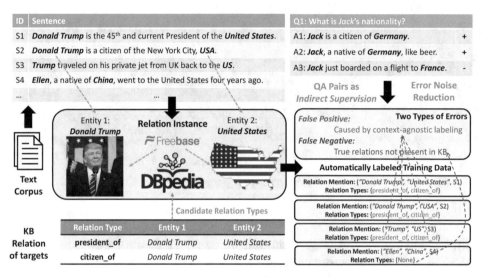

Figure 10.1: Distant supervision generates training data by linking relation mentions in sentences S1–S4 to KB and assigning the linkable relation types to all relation mentions. Those unlinkable entity mention pairs are treated as negative examples. This automatic labeling process may cause errors of false positives (highlighted in red) and false negatives (highlighted in purple). QA pairs provide indirect supervision for correcting such errors.

10.1.1 CHALLENGES

Toward the goal of diminishing the negative effects by noisy DS training data, distantly supervised RE models that deal with training noise, as well as methods that directly improve the automatic training data generation process have been proposed. These methods mostly involve designing distinct assumptions to remove redundant training information [Hoffmann et al., 2011, Lin et al., 2016, Mintz et al., 2009, Riedel et al., 2010]. For example, the method applied in Hoffmann et al. [2011] and Riedel et al. [2010], assumes that for each relation triple in the KB, at least one sentence might express the relation instead of all sentences. Moreover, these noise reduction systems usually only address one type of error, either false positives or false negatives. Hence, current methods handling DS noises still have the following challenges.

1. Lack of trustworthy sources: Current de-noising methods mainly focus on recognizing labeling mistakes from the labeled data itself, assisted by pre-defined assumptions or patterns. They do not have external trustworthy sources as guidance to uncover incorrectly

labeled data, while not at the expense of excessive human efforts. Without other separate information sources, the reliability of false label identification can be limited.

2. Incomplete noise handling: Although both false negative and false positive errors are observed to be significant, most existing works only address one of them.

10.1.2 PROPOSED SOLUTION

In this chapter, to overcome the above two issues derived from relation extraction with distant supervision, we study the problem of relation extraction with indirect supervision from external sources. Recently, the rapid emergence of QA systems promotes the availability of user feedback or datasets of various QA tasks. We investigate to leverage QA, a downstream application of relation extraction, to provide additional signals for learning RE models. Specifically, we use datasets for the task of answer sentence selection to facilitate relation typing. Given a domain-specific corpus and a set of target relation types from a KB, we aim to detect relation mentions from text and categorize each in context by target types or Non-Target-Type (None) by leveraging an independent dataset of QA pairs in the same domain. We address the above two challenges as follows. (1) We integrate indirect supervision from another same-domain data source in the format of QA sentence pairs, that is, each question sentence maps to several positive (where a true answer can be found) and negative (where no answer exists) answer sentences. We adopt the principle that for the same question, positive pairs of (question, answer) should be semantically similar while they should be dissimilar from negative pairs. (2) Instead of differentiating types of labeling errors at the instance level, we concentrate on how to better learn semantic representation of features. Wrongly labeled training examples essentially misguide the understanding of features. It increases the risk of having a non-representative feature learned to be close to a relation type and vice versa. Therefore, if the feature learning process is improved, potentially both types of error can be reduced.

To integrate all the above elements, a novel framework, ReQuest, is proposed. First, ReQuest constructs a heterogeneous graph to represent three kinds of objects: relation mentions, text features and relation types for RE training data labeled by KB linking. Then, ReQuest constructs a second heterogeneous graph to represent entity mention pairs (include question, answer entity mention pairs) and features for QA dataset. These two graphs are combined into a single graph by overlapped features. We formulate a global objective to jointly embed the graph into a low-dimensional space where, in that space, RE objects whose types are semantically close also have similar representations and QA objects linked by positive (question, answer) entity mention pairs of a same question should have close representations. In particular, we design a novel margin-based loss to model the semantic similarity between QA pairs and transmit such information into feature and relation type representations via shared features. With the learned embeddings, we can efficiently estimate the types for test relation mentions.

10.2 THE PROPOSED APPROACH

Framework Overview. We propose an *embedding-based* framework with indirect supervision (illustrated in Fig. 10.2) as follows.

1. Generate text features for each relation mention or QA entity mention pair, and construct a heterogeneous graph using four kinds of objects in combined corpus, namely relation mentions from RE corpus, entity mention pairs from QA corpus, target relation types, and text features to encode aforementioned signals in a unified form (Section 10.2.1).

2. Jointly embed relation mentions, QA pairs, text features, and type labels into two low-dimensional spaces connected by shared features, where close objects tend to share the same types or questions (Section 10.2.2).

3. Estimate type labels r^* for each test relation mention z from learned embeddings, by searching the target type set $\mathbb{R}$ (Section 10.2.3).

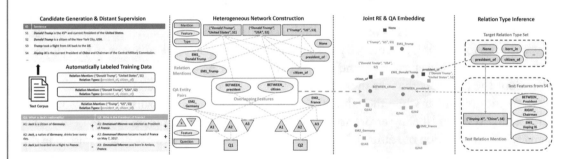

Figure 10.2: Overall framework.

10.2.1 HETEROGENEOUS NETWORK CONSTRUCTION

Relation Mentions and Types Generation. We get the relation mentions along with their heuristically obtained relation types from the automatically labeled training corpus $\mathcal{D}_L$. And we randomly sample a set of unlinkable entity mention pairs as the negative relation mentions (i.e., relation mentions assigned with type "None").

QA Entity Mention Pairs Generation. We apply Stanford NER [Manning et al., 2014] to extract entity mentions in each question or answer sentence. First, we detect the target entity being asked about in each question sentence. For example, in the question "*Who is the president of United States,*" the question entity is "*United States.*" In most cases, a question only contains one entity mention and for those containing multiple entity mentions, we notice the question entity is mostly mentioned at the very last. Thus, we follow this heuristic rule to assign the lastly occurred entity mention to be the question entity mention m_0 in each question sentence q_i.

Then, in each positive answer sentence of q_i, we extract the entity mention with matched head token and smallest edit string distance to be the question entity mention m_1, and the entity mention matching the exact answer string to be the answer entity mention m_2. Then we form a positive QA entity mention pair with its context s, $p_k = (m_1, m_2, s) \in \mathcal{P}_i^+$ for q_i. If either m_1 or m_2 cannot be found, this positive answer sentence is dropped. We randomly select pairs of entity mentions in each negative answer sentence to be negative QA entity mention pairs for q_i (e.g., if a negative sentence includes 3 entity mentions, we randomly select negative examples from the $3 \cdot 2 \cdot 1 = 6$ different pairs of entity mentions in total, if we ignore the order), with each negative example marked as $p_{k\prime} = (m_{1\prime}, m_{2\prime}, s\prime) \in \mathcal{P}_i^-$ for q_i.

Text Feature Extraction. We extract lexical features of various types from not only the mention itself (e.g., head token), as well as the context s (e.g., bigram) in a POS-tagged corpus. It is to capture the syntactic and semantic information for any given relation mentions or entity mention pairs. We follow those used in Chan and Roth [2010] and Mintz et al. [2009] (excluding the dependency parse-based features and entity type features).

We denote the set of M_z unique features extracted from relation mentions $\mathcal{Z}$ as $\mathcal{F}_z = \{f_j\}_{j=1}^{M_z}$ and the set of M_{QA} unique features extracted of QA entity mention pairs $\mathcal{P}$ as $\mathcal{F}_{QA} = \{f_j\}_{j=1}^{M_{QA}}$. As our embedding learning process will combine these two sets of features and their shared ones will act as the bridge of two embedding spaces, we denote the overall feature set as $\mathcal{F} = \{f_j\}_{j=1}^{M}$.

Heterogeneous Network Construction. After the nodes generation process, we construct a heterogeneous network connected by text features, relation mentions, relation types, questions, QA entity mention pairs, as shown in the second column of Fig. 10.2.

10.2.2 JOINT RE AND QA EMBEDDING

This section first introduces how we model different types of interactions between linkable relation mentions $\mathcal{Z}$, QA entity mention pairs $\mathcal{P}$, relation type labels $\mathbb{R}$ and text features $\mathcal{F}$ into a *d-dimensional relation vector space* and a d-dimensional *QA pair vector space*. In the relation vector space, objects whose types are close to each other should have similar representation and in the QA pair vector space, positive QA mention pairs who share the same question are close to each other. (e.g., see the third column in Fig. 10.2). We then combine multiple objectives and formulate a joint optimization problem.

We propose a novel global objective, which employs a margin-based rank loss [Nguyen and Caruana, 2008] to model *noisy mention-type associations* and utilizes the second-order proximity idea [Tang et al., 2015] to model *mention-feature (QA pair-feature) co-occurrences*. In particular, we adopt a pairwise margin loss, following the intuition of pairwise rank [Rao et al., 2016] to capture the *interactions between QA pairs*, and the shared features $\mathcal{F}_z \cap \mathcal{F}_{QA}$ between relation mentions $\mathcal{Z}$ and QA pairs $\mathcal{P}$ connect the two vector spaces.

A Joint Optimization Problem. Our goal is to embed all the available information for relation mentions and relation types, QA entity mention pairs and text features into a single d-

dimensional embedding space. An intuitive solution is to collectively minimize the two objectives $\mathcal{O}_Z$ and $\mathcal{O}_{QA}$ as the embedding vectors of overlapped text features are shared across relation vector space and QA pair vector space. To achieve the goal, we formulate a joint optimization problem as follows:

$$\min \mathcal{O} = \mathcal{O}_Z + \mathcal{O}_{QA}. \tag{10.1}$$

When optimizing the global objective O, the learning of RE and QA embeddings can be mutually influenced as errors in each component can be constrained and corrected by the other. This mutual enhancement also helps better learn the semantic relations between features and relation types. We apply edge sampling strategy [Tang et al., 2015] with a stochastic sub-gradient descent algorithm [Shalev-Shwartz et al., 2011] to efficiently solve Eq. (10.1). In each iteration, we alternatively sample from each of the two objectives $\{\mathcal{O}_Z, \mathcal{O}_M\}$ a batch of edges (e.g., (z_i, f_j)) and their negative samples, and update each embedding vector based on the derivatives.

10.2.3 TYPE INFERENCE

To predict the type for each test relation mention z, we search for nearest neighbor in the target relation type set $\mathbb{R}$, with the learned embeddings of features and relation types. Specifically, we represent test relation mention z in our learned relation embedding space by $\mathbf{t}z = \sum_{f_j \in \mathcal{F}_z(z)} \mathbf{t}c_j$ where $\mathcal{F}_z(z)$ is the set of text features extracted from z's local context s. We categorize z to None type if the similarity score is below a pre-defined threshold (e.g., $\eta > 0$).

10.3 EXPERIMENTS

Data sets. Our experiments consists of two different type of datasets, one for relation extraction and another answer sentence selection dataset for indirect supervision. Two public datasets are used for relation extraction: **NYT** [Hoffmann et al., 2011, Riedel et al., 2010] and **KBP** [Ellis et al., 2014, Ling and Weld, 2012]. The test data are manually annotated with relation types by their respective authors.

Compared Methods. We compare REQUEST with its variants which model parts of the proposed hypotheses. Several state-of-the-art relation extraction methods (e.g., supervised, embedding, neural network) are also implemented (or tested using their published codes). Besides the proposed joint optimization model, **ReQuest-Joint**, we conduct experiments on two other variations to compare the performance (1) **ReQuest-QA_RE**: This variation optimizes objective $\mathcal{O}_{QA}$ first and then uses the learned feature embeddings as the initial state to optimize $\mathcal{O}_Z$; and (2) **ReQuest-RE_QA**: It first optimizes $\mathcal{O}_Z$, then optimizes $\mathcal{O}_{QA}$ to finely tune the learned feature embeddings.

Performance Comparison with Baselines. To test the effectiveness of our proposed framework REQUEST, we compare with other methods on the relation extraction task. The precision, recall, F1 scores as well as the model learning time measured on two datasets are reported in Table 10.1. As shown in the table, REQUEST achieves superior F1 score on both datasets compared with

Table 10.1: Performance comparison on end-to-end relation extraction (at the highest F1 point) on the two datasets

Method	NYT [Riedel et al., Hoffmann et al., 2011]				KBP [Ling and Weld, 2012, Ellis et al., 2014]			
	Prec	Rec	F1	Time	Prec	Rec	F1	Time
DS+Perceptron [Ling and Weld, 2012]	0.068	**0.641**	0.123	15 min	0.233	0.457	0.308	7.7 min
DS+Kernel [Mooney and Bunescu, 2016]	0.095	0.490	0.158	56 hr	0.108	0.239	0.149	9.8 hr
DS+Logistic [Mintz et al., 2009]	0.258	0.393	0.311	25 min	0.296	0.387	0.335	14 min
DeepWalk [Perozzi et al., 2014]	0.176	0.224	0.197	1.1 hr	0.101	0.296	0.150	27 min
LINE [Tang et al., 2015]	0.335	0.329	0.332	2.3 min	0.360	0.257	0.299	1.5 min
MultiR [Hoffmann et al., 2011]	0.338	0.327	0.333	5.8 min	0.325	0.278	0.301	4.1 min
FCM [Gourmley et al., 2015]	**0.553**	0.154	0.240	1.3 hr	0.151	**0.500**	0.301	25 min
DS+SDP-LSTM [Xu et al., 2015]	0.307	0.532	0.389	21 hr	0.249	0.300	0.272	10 hr
DS+LSTM-ER [Miwa and Bansal, 2016]	0.373	0.171	0.234	49 hr	0.338	0.106	0.161	30 hr
CoType-RM [Ren et al., 2017]	0.467	0.380	0.419	2.6 min	0.342	0.339	0.340	1.5 min
ReQuest-QA_RE	0.407	0.437	0.422	10.2 min	**0.459**	0.300	0.363	5.3 min
ReQuest-RE_QA	0.435	0.419	0.427	8.0 min	0.356	0.352	0.354	13.2 min
ReQuest-Joint	0.404	0.480	**0.439**	4.0 min	0.386	0.410	**0.397**	5.9 min

other models. Among all these baselines, MultiR and CoType-RM handle noisy training data while the remaining ones assume the training corpus is perfectly labeled. Due to their nature of being cautious toward the noisy training data, both MultiR and CoType-RM reach relatively high results confronting with other models that blindly exploit all heuristically obtained training examples. However, as external reliable information sources are absent and only the noise from multi-label relation mentions (while none or only one assigned label is correct) is tackled in these models, MultiR and CoType-RM underperform ReQuest. Especially from the comparison with CoType-RM, which is also an embedding learning based relation extraction model with the idea of partial-label loss incorporated, we can conclude that the extra semantic inklings provided by the QA corpus do help boost the performance of relation extraction.

Performance Comparison with Ablations. We experiment with two variations of ReQuest, ReQuest-QA_RE, and ReQuest-RE_QA, in order to validate the idea of joint optimization.

As presented in Table 10.1, both REQUEST-QA_RE and REQUEST-RE_QA outperform most of the baselines, with the indirect supervision from QA corpus. However, their results still fall behind REQUEST's. Thus, separately training the two components may not capture as much information as jointly optimizing the combined objective. The idea of constraining each component in the joint optimization process proves to be effective in learning embeddings to present semantic meanings of objects (e.g., features, types and mentions).

10.4   SUMMARY

We present a novel study on indirect supervision (from question-answering datasets) for the task of relation extraction. We propose a framework, REQUEST, that embeds information from both training data automatically generated by linking to KBs and QA datasets, and captures richer semantic knowledge from both sources via shared text features so that better feature embeddings can be learned to infer relation type for test relation mentions despite the noisy training data. Our experiment results on two datasets demonstrate the effectiveness and robustness of REQUEST. Interesting future work includes identifying most relevant QA pairs for target relation types, generating most effective questions to collect feedback (or answers) via crowd-sourcing, and exploring approaches other than distant supervision [Artzi and Zettlemoyer, 2013, Riedel et al., 2013].

PART III

Toward Automated Factual Structure Mining

CHAPTER 11

Mining Entity Attribute Values with Meta Patterns

Meng Jiang, *Department of Computer Science and Engineering, University of Notre Dame*

Discovering *textual patterns* from text data is an active research theme [Banko et al., 2007, Carlson et al., 2010, Fader et al., 2011, Gupta et al., 2014, Nakashole et al., 2012], with broad applications such as attribute extraction [Ghani et al., 2006, Pasca, 2008, Probst et al., 2007, Ravi and Paşca, 2008], aspect mining [Chen et al., 2014, Hu and Liu, 2014, Kannan et al., 2011], and slot filling [Yahya et al., 2014, Yu and Ji, 2016]. Moreover, a data-driven exploration of *efficient* textual pattern mining may also have strong implications on the development of efficient methods for NLP tasks on massive text corpora.

11.1 OVERVIEW AND MOTIVATION

Traditional methods of textual pattern mining have made large pattern collections publicly available, but very few can extract arbitrary patterns with semantic types. Hearst patterns like "*NP* such as *NP*, *NP*, and *NP*" were proposed and widely used to acquire hyponymy lexical relation [Hearst, 1992]. TextRunner [Banko et al., 2007] and ReVerb [Fader et al., 2011] are blind to the typing information in their lexical patterns; ReVerb constrains patterns to verbs or verb phrases that end with prepositions. NELL [Carlson et al., 2010] learns to extract noun-phrase pairs based on a fixed set of prespecified relations with entity types like country:president→$COUNTRY×$POLITICIAN.

 One interesting exception is the SOL patterns proposed by Nakashole et al. [2012] in PATTY. PATTY relies on the Stanford dependency parser [Marneffe et al., 2006] and harnesses the typing information from a KB [Auer et al., 2007, Bollacker et al., 2008, Nastase et al., 2010] or a typing system [Ling and Weld, 2012, Nakashole et al., 2013]. Despite of the significant contributions of the work, SOL patterns have three limitations on mining typed textual patterns from a large-scale text corpus, as illustrated below.

11.1.1 CHALLENGES

First, a good typed textual pattern should be of informative, self-contained context. The dependency parsing in PATTY loses the rich context around the entities such as the word "president" next to "Barack Obama" in sentence #1, and "president" and "prime_minister" in #2. Moreover, the SOL patterns are restricted to the dependency path between two entities but do not represent the data types like $DIGIT for "55" and $MONTH $DAY $YEAR. Furthermore, the parsing process is costly: Its complexity is cubic in the length of sentence [McDonald et al., 2005], which is too costly for news and scientific corpora that often have long sentences. We expect an efficient textual pattern mining method for massive corpora.

Second, synonymous textual patterns are expected to be identified and grouped for handling pattern sparseness and aggregating their extractions for extending KBs and question answering. However, the process of finding such synonymous pattern groups is non-trivial. Multi-faceted information should be considered: (1) synonymous patterns should share the same entity types or data types; (2) even for the same entity (e.g., Barack Obama), one should allow it be grouped and generalized differently (e.g., in ⟨United States, Barack Obama⟩ vs. ⟨Barack Obama, 55⟩); and (3) shared words (e.g., "president") or semantically similar contextual words (e.g., "age" and "-year-old") may play an important role in synonymous pattern grouping. PATTY does not explore the multi-faceted information at grouping syonymous patterns, and thus cannot aggregate such extractions.

Third, the entity types in the textual patterns should be precise. In different patterns, even the same entity can be typed at different type levels. For example, the entity "Barack Obama" should be typed at a fine-grained level ($POLITICIAN) in the patterns generated from sentence #1–2, and it should be typed at a coarse-grained level ($PERSON) in the patterns from sentence #3–4. However, PATTY does not look for appropriate granularity of the entity types.

11.1.2 PROPOSED SOLUTION

In this chapter, we propose a new typed textual pattern called *meta pattern*, which is defined as follows.

Definition 11.1 Meta Pattern.   A meta pattern refers to a frequent, informative, and precise subsequence pattern of entity types (e.g., $PERSON, $POLITICIAN, $COUNTRY) or data types (e.g., $DIGIT, $MONTH, $YEAR), words (e.g., "politician," "age"), or phrases (e.g., "prime_minister"), and possibly punctuation marks (e.g., ",", "("), which serves as an integral semantic unit in certain context.

We study the problem of mining meta patterns and grouping synonymous meta patterns. *Why mining meta patterns and grouping them into synonymous meta pattern groups?*—because mining and grouping meta patterns into synonymous groups may facilitate information extraction and turning unstructured data into structures. For example, given us a sentence from a news corpus, "President Blaise Compaoré's government of Burkina Faso was founded ...", if we have discovered the meta pattern "president \$POLITICIAN's government of \$COUNTRY," we can recognize and type new entities (i.e., type "Blaise Compaoré" as a \$POLITICIAN and "Burkina Faso" as a \$COUNTRY), which previously requires human expertise on language rules or heavy annotations for learning [Nadeau and Sekine, 2007]. If we have grouped the pattern with synonymous patterns like "\$COUNTRY president \$POLITICIAN", we can merge the fact tuple ⟨Burkina Faso, president, Blaise Compaoré ⟩ into the large collection of facts of the attribute type country:president.

To systematically address the challenges of mining meta patterns and grouping synonymous patterns, we develop a novel framework called MetaPAD (Meta PAttern Discovery). Instead of working on every individual sentence, our MetaPAD leverages *massive* sentences in which redundant patterns are used to express attributes or relations of *massive* instances. First, MetaPAD generates meta pattern candidates using efficient sequential pattern mining, learns a quality assessment function of the patterns candidates with a rich set of domain-independent contextual features for intuitive ideas (e.g., frequency, informativeness), and then mines the quality meta patterns by assessment-led context-aware segmentation (see Section 11.2.1). Second, MetaPAD formulates the grouping process of synonymous meta patterns as a learning task, and solves it by integrating features from multiple facets including entity types, data types, pattern context, and extracted instances (see Section 11.2.2). Third, MetaPAD examines the type distributions of entities in the extractions from every meta pattern group, and looks for the most appropriate type level that the patterns fit. This includes both top-down and bottom-up schemes that traverse the type ontology for the patterns' preciseness (see Section 11.2.3).

11.1.3 PROBLEM FORMULATION

Problem 11.2   Meta Pattern Discovery   Given a fine-grained, typed corpus of massive sentences $C = [\ldots, S, \ldots]$, and each sentence is denoted as $S = t_1 t_2 \ldots t_n$ in which $t_k \in \mathcal{T} \cup \mathcal{P} \cup \mathcal{M}$ is the k-th token ($\mathcal{T}$ is the set of entity types and data types, $\mathcal{P}$ is the set of phrases and words, and $\mathcal{M}$ is the set of punctuation marks), the task is to find **synonymous groups of quality meta patterns**. A *meta pattern mp* is a subsequential pattern of the tokens from the set $\mathcal{T} \cup \mathcal{P} \cup \mathcal{M}$. A *synonymous meta pattern group* is denoted by $\mathcal{MPG} = [\ldots, mp_i, \ldots, mp_j \ldots]$ in which each pair of meta patterns, mp_i and mp_j, are synonymous.

What is a quality meta pattern? Here we take the sentences as sequences of tokens. Previous sequential pattern mining algorithms mine frequent subsequences satisfying a single metric, the minimum support threshold (min_sup), in a transactional sequence database [Agrawal and Srikant, 1995]. However, for text sequence data, the quality of our proposed textual pattern, the meta pattern, should be evaluated similar to phrase mining [Liu et al., 2015] in four criteria, as illustrated below.

Example 11.3   The quality of a pattern is evaluated with the following criteria (the former pattern has higher quality than the latter).

1. *Frequency:* "$DIGITRANK president of $COUNTRY" vs. "young president of $COUNTRY."

2. *Completeness:* "$COUNTRY president $POLITICIAN" vs. "$COUNTRY president," "$PERSON's wife, $PERSON" vs. "$PERSON's wife."

3. *Informativeness:* "$PERSON's wife, $PERSON" vs. "$PERSON and $PERSON."

4. *Preciseness:* "$COUNTRY president $POLITICIAN" vs. "$LOCATION president $PERSON," "$PERSON's wife, $PERSON" vs. "$POLITICIAN's wife, $PERSON," "population of $LOCATION" vs. "population of $COUNTRY."

What are synonymous meta patterns? The full set of frequent sequential patterns from a transaction dataset is huge [Agrawal and Srikant, 1995]; and the number of meta patterns from a massive corpus is also big. Since there are multiple ways to express the same or similar meanings in a natural language, many meta patterns may share the same or nearly the same meaning. Grouping synonymous meta patterns can help aggregate a large number of extractions of different patterns from different sentences. And the type distribution of the aggregated extractions can help us adjust the meta patterns in the group for preciseness.

11.2   THE METAPAD FRAMEWORK

Figure 11.1 presents the MetaPAD framework for Meta PAttern Discovery. It has three modules. First, it develops a context-aware segmentation method to determine the boundaries of the sub-sequences and generate the meta patterns of frequency, completeness, and informativeness (see Section 11.2.1). Second, it groups synonymous meta patterns into clusters (see Section 11.2.2). Third, for every synonymous pattern group, it adjusts the levels of entity types for appropriate granularity to have precise meta patterns (see Section 11.2.3).

11.2.1   GENERATING META PATTERNS BY CONTEXT-AWARE SEGMENTATION

Pattern candidate generation.   We adopt the standard frequent sequential pattern mining algorithm [Pei et al., 2004] to look for pattern candidates that satisfy a min_sup threshold. In practice, one can set a maximum pattern length ω to restrict the number of tokens in the patterns. Different from syntactic analysis of very long sentences, our meta pattern mining explores pattern structures that are local but still of wide context: in our experiments, we set $\omega = 20$.

Meta pattern quality assessment.   Given a huge number of pattern candidates that can be messy (e.g., "of $COUNTRY" and "$POLITICIAN and"), it is desired but challenging to assess the quality of the patterns with a very few training labels. We introduce a rich set of *contextual*

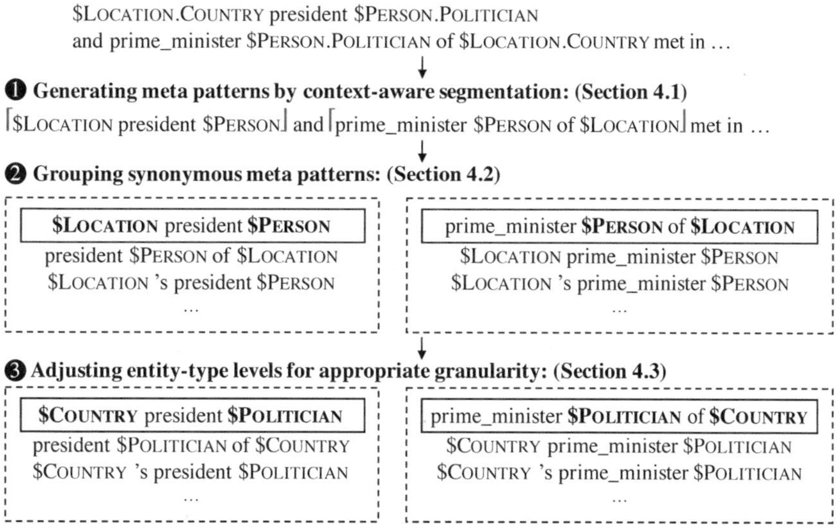

$LOCATION.COUNTRY president $PERSON.POLITICIAN
and prime_minister $PERSON.POLITICIAN of $LOCATION.COUNTRY met in …

↓

❶ **Generating meta patterns by context-aware segmentation: (Section 4.1)**

⌈$LOCATION president $PERSON⌋ and ⌈prime_minister $PERSON of $LOCATION⌋ met in …

↓

❷ **Grouping synonymous meta patterns: (Section 4.2)**

$LOCATION president $PERSON
president $PERSON of $LOCATION
$LOCATION 's president $PERSON
…

prime_minister $PERSON of $LOCATION
$LOCATION prime_minister $PERSON
$LOCATION 's prime_minister $PERSON
…

↓

❸ **Adjusting entity-type levels for appropriate granularity: (Section 4.3)**

$COUNTRY president $POLITICIAN
president $POLITICIAN of $COUNTRY
$COUNTRY 's president $POLITICIAN
…

prime_minister $POLITICIAN of $COUNTRY
$COUNTRY prime_minister $POLITICIAN
$COUNTRY 's prime_minister $POLITICIAN
…

Figure 11.1: Three modules in our MetaPAD framework.

features of the patterns according to the quality criteria (see Section 11.1.3) and train a classifier to estimate the quality function $Q(mp) \in [0, 1]$ where mp is a meta pattern candidate.

We train a classifier based on random forests [Breiman, 2001] for learning the meta-pattern quality function $Q(mp)$ with the above rich set of contextual features. Our experiments (not reported here for the sake of space) show that using only 100 pattern labels can achieve similar precision and recall as using 300 labels. Note that the learning results can be transferred to other domains: the features of low-quality patterns "$POLITICIAN and $COUNTRY" and "$BACTERIA and $ANTIBIOTICS" are similar; the features of high-quality patterns "$POLITICIAN is president of $COUNTRY" and "$BACTERIA is resistant to $ANTIBIOTICS" are similar.

Context-aware segmentation using $Q(.)$ with feedback.   With the pattern quality function $Q(.)$ learned from the rich set of contextual features, we develop a bottom-up segmentation algorithm to construct the best partition of segments of high quality scores. As shown in Fig. 11.2, we use $Q(.)$ to determine the boundaries of the segments: we take "$COUNTRY president $POLITICIAN" for its high quality score; we do not take the candidate "and prime_minister $POLITICIAN of $COUNTRY" because of its low quality score.

Since $Q(mp)$ was learned with features including the raw frequency $c(mp)$, the quality score may be overestimated or underestimated: the principle is that every token's occurrence should be assigned to only one pattern but the raw frequency may count the tokens multiple times. Fortunately, after the segmentation, we can rectify the frequency as $c_r(mp)$, for example in Fig. 11.2, the segmentation avoids counting "$POLITICIAN and prime_minister $POLITICIAN" of overestimated frequency/quality. Once the frequency feature is rectified, we re-learn the quality

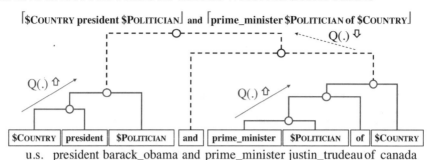

Figure 11.2: Generating meta patterns by context-aware segmentation with the pattern quality function $Q(.)$.

function $Q(.)$ using $c(mp)$ as feedback and re-segment the corpus with it. This can be an iterative process but we found in only one iteration, the result converges.

11.2.2 GROUPING SYNONYMOUS META PATTERNS

Grouping truly synonymous meta patterns enables a large collection of extractions of the same relation aggregated from different but synonymous patterns. For example, there could be hundreds of ways of expressing the relation country:president; if we group all such meta patterns, we can aggregate all the extractions of this relation from massive corpus. PATTY [Nakashole et al., 2012] has a narrow definition of their synonymous dependency path-based SOL patterns: two patterns are synonymous if they generate the same set of extractions from the corpus. Here we develop a learning method to incorporate information of three aspects: (1) entity/data types in the pattern, (2) context words/phrases in the pattern, and (3) extractions from the pattern, to assign the meta patterns into groups. Our method is based on three assumptions as follows.

A1: Synonymous meta patterns must have the same entity/data types: the meta patterns "$PERSON's age is $DIGIT" and "$PERSON's wife is $PERSON" cannot be synonymous.

A2: If two meta patterns share (nearly) the same context words/phrases, they are more likely to be synonymous: the patterns "$COUNTRY president $POLITICIAN" and "president $POLITICIAN of $COUNTRY" share the word "president."

A3: If two patterns generate more common extractions, they are more likely to be synonymous: both "$PERSON's age is $DIGIT" and "$PERSON, $DIGIT," generate ⟨Barack Obama, 55⟩.

Since the number of groups cannot be pre-specified, we propose to first construct a pattern-pattern graph in which the two pattern nodes of every edge satisfy *A1* and are predicted to be synonymous, and then use a clique detection technique [Harary and Ross, 1957] to find all the cliques as synonymous meta patten groups. Each pair of the patterns (mp_i, mp_j) in the group $\mathcal{MPG} = [\ldots, mp_i, \ldots, mp_j \ldots]$ are synonymous.

For the graph construction, we train Support Vector Regression Machines [Drucker et al., 1997] to learn the following features of a pair of patterns based on *A2* and *A3*: (1) the numbers of words, non-stop words, phrases that each pattern has and they share; (2) the maximum similarity score between pairs of non-stop words or phrases in the two patterns; and (3) the number of extractions that each pattern has and they share. The similarity between words/phrases is represented by the cosine similarity of their word2vec embeddings [Mikolov et al., 2013, Toutanova et al., 2015].

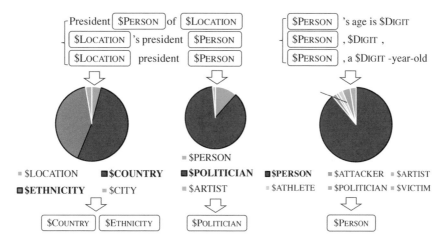

Figure 11.3: Adjusting entity type levels for appropriate granularity with entity type distributions.

11.2.3 ADJUSTING TYPE LEVELS FOR PRECISENESS

Given a group of synonymous meta patterns, we expect the patterns to be precise: it is desired to determine the levels of the entity types in the patterns for appropriate granularity. Thanks to the grouping process of synonymous meta patterns, we have rich type distributions of the entities from the large collection of extractions.

11.3 SUMMARY

In this chapter, we proposed a novel typed textual pattern structure, called *meta pattern*, which is extended to a frequent, complete, informative, and precise subsequence pattern in certain context, compared with the SOL pattern. We developed an efficient framework, MetaPAD, to discover the meta patterns from massive corpora with three techniques, including: (1) a context-aware segmentation method to carefully determine the boundaries of the patterns with a learned pattern quality assessment function, which avoids costly dependency parsing and generates high-quality patterns; (2) a clustering method to group synonymous meta patterns with integrated information of types, context, and instances; and (3) top-down and bottom-up schemes to adjust the levels of entity types in the meta patterns by examining the type distributions of entities in the instances. Experiments demonstrated that MetaPAD efficiently discovered a large collection of high-quality typed textual patterns to facilitate challenging NLP tasks like tuple information extraction.

CHAPTER 12

Open Information Extraction with Global Structure Cohesiveness

Qi Zhu, *Department of Computer Science, University of Illinois at Urbana-Champaign*

Massive text corpora are emerging worldwide in different domains and languages. The sheer size of such unstructured data and the rapid growth of new data pose grand challenges on making sense of these massive corpora. Information extraction (IE) [Sarawagi, 2008]—extraction of relation tuples in the form of (*head entity*, relation, *tail entity*)—is a key step toward automating knowledge acquisition from text. While traditional IE systems require people to pre-specify the set of relations of interests, recent studies on *open-domain information extraction* (Open IE) [Banko et al., 2007, Carlson et al., 2010, Schmitz et al., 2012] rely on *relation phrases* extracted from text to represent the entity relationship, making it possible to adapt to various domains (i.e., open-domain) and different languages (i.e., language-independent).

12.1 OVERVIEW AND MOTIVATION

Prior work on Open IE can be summarized as sharing two common characteristics: (1) conducting extraction based on local context information; and (2) adopting an incremental system pipeline. Current Open IE systems focus on analyzing the local context within individual sentences to extract entity and their relationships, while ignoring the redundant information that can be collectively referenced across different sentences and documents in the corpus. For example, in Fig. 12.1, seeing entity phrases "*London*" and "*Paris*" frequently co-occur with similar relation phrase and tail entities in the corpus, one gets to know that they have close semantics (same for "*Great Britain*" and "*France*"). On one hand, this helps confirm that (*Paris*, is in, *France*) is a quality tuple if knowing (*London*, is in , *Great Britain*) is a good tuple. On the other, this helps rule out the tuple (*Paris*, build, *new satellites*) as "*Louvre-Lens*" is semantically distant from "*Paris*." Therefore, the rich information redundancy in the massive corpus motivates us to design an effective way of measuring whether a candidate relation tuple is consistently used across various context in the corpus (i.e., global cohesiveness).

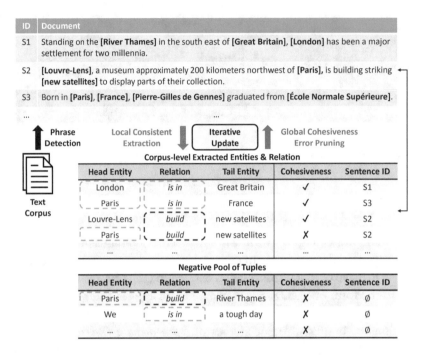

Figure 12.1: Example of incorporating global cohesiveness view for error pruning. "London" and "Paris" are similar because they are head entities of the same relation "is in." When it comes to the relation "build," since "London" and "build" do not co-occur in any tuple in the corpus, then it is unlikely for tuples with "Paris" and "build" to be correct.

Furthermore, most existing Open IE systems assume that they have access to entity detection tools (e.g., named entity recognizer (NER), noun phrase (NP) chunker) to extracted entity phrases from sentences, which are then used to form entity pairs for relation tuple extraction [Banko et al., 2007, Carlson et al., 2010, Schmitz et al., 2012]. Some systems further rely on dependency parsers to generate syntax parse tree for guiding the relation tuple extraction [Angeli et al., 2015, Corro and Gemulla, 2013, Schmitz et al., 2012]. However, these systems suffer from *error propagation* as the errors in prior parts of the pipeline could accumulate cascading down the pipeline, yielding more significant errors. In addition, the NERs and NP chunkers are often pre-trained for general domain and may not work well on a domain-specific corpus (e.g., scientific papers, social media posts).

12.1.1 PROPOSED SOLUTION

In this chapter, we propose a novel framework, called ReMine, to unify two important yet *complementary* signals for Open IE problem, i.e., the local context information and the global cohesiveness (see also Fig. 12.2). While most existing Open IE systems focus on analyzing local context and linguistic structures for tuple extraction, ReMine further make use of all the candidate tuples extracted from the entire corpus, to collectively measure whether these candidate tuples are reflecting cohesive semantics. This is done by mapping both entity and relation phrases into the same low-dimensional embeddings space, where two entity phrases are similar if they share similar relation phrases and entity arguments. The entity and relation embeddings so learned can be used to measure the cohesiveness score of a candidate relation tuple. To overcome the error propagation issue, ReMine *jointly* optimizes both the *extraction of entity and relation phrases* and the *global cohesiveness across the corpus*, each being formalized as an objective function so as to quantify the quality scores, respectively.

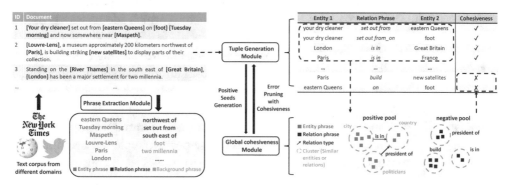

Figure 12.2: Overview of the ReMine framework.

Specifically, ReMine first identifies entity and relation phrases from local context. In Fig. 12.2, suppose we have a sentence "Your dry cleaner set out from eastern Queens on foot Tuesday morning and now somewhere near Maspeth." We will first extract three entity phrases, *eastern Queens*, *Tuesday morning*, *Maspeth*, as well as two background phrases *Your dry cleaner*, *foot*. Then, ReMine jointly mines relation tuples and measure extraction with global translating objective. Local consistent text segmentation may generate noisy tuples, such as <your dry cleaner, *set out from*, eastern Queens> and <eastern Queens, *on*, foot>. However, from the global cohesiveness view, we may infer the second tuple as a false positive. Entity phrases like "eastern Queens" are seldom linked by relation phrase "*on*" in extracted tuples. Overall, ReMine will iteratively refine extracted tuples and learn entity and relation representations from corpus level. With careful attention to advantages of linguistic patterns [Fader et al., 2011, Hearst, 1992] and representation learning [Bordes et al., 2013], this approach benefits from both side. Compared to previous open IE systems, ReMine prune extracted tuples via global cohesiveness and its accuracy is not sensitive to the target domain.

12.2   THE REMINE FRAMEWORK

ReMine aims to jointly address two problems, that is, extracting entity and relation phrases from sentences and generating quality relation tuples. There are three challenges. First, distant supervision may contain *false* seed examples of entity and relation phrases, and thus asks for effective measuring of the quality score for phrase candidates. Second, there exist multiple entity phrases in one sentence. Therefore, selecting entities to form relation tuples may suffer from ambiguity in local context. Third, ranking extracted tuples without referring to the entire corpus may favor with good local structures.

Framework Overview. We proposed a framework, called ReMine, that integrates both local context and global structure cohesiveness (see also Fig. 12.2) to address above challenges. There are three major modules in ReMine: (1) phrase extraction module; (2) relation tuple generation; and (3) global cohesiveness module. To overcome sparse and noisy labels, phrase extraction module trains a robust phrase classifier and adjusts quality from a generative perspective. The relation tuple extraction module generates tuples from sentence structure, which adopts widely used local structure patterns [Corro and Gemulla, 2013, Nakashole et al., 2012, Schmitz et al., 2012], including syntactic and lexically patterns over pos tags and dependency parsing tree. However different from previous studies, the module tries to benefit from information redundancy and mine distinctive extractions with accurate relation phrases. Meanwhile, global cohesiveness module learns entity and relation phrases representation with a score function to rank tuples. Relation tuple generation module and global cohesiveness module are collaborating with each other. Particularly, relation tuple generation module produces coarse positive tuple seeds and feeds them into global cohesiveness module. By distinguishing positive tuples with constructed negative samples, global cohesiveness module provide a cohesiveness measure for tuple generation. Tuple generator further incorporates global cohesiveness into local generation and outputs more precise extractions. ReMine integrates tuple generation and global cohesiveness learning into a joint optimization framework. They iteratively refine input for each other and eventually obtain clean extractions. Once the training process converges, the tuples are expected to be distinctive and accurate. Overall, ReMine extracts relation tuples as follows.

1. **Phrase extraction module** conducts context-dependent phrasal segmentation on target corpus (using distant supervision) , to generate entity phrases $\mathcal{E}$, relation phrases $\mathbb{R}$, and word sequence probability $\mathcal{A}$.

2. **Relation tuple generation module** generates positive entity pairs and identifies predicate p between entity argument pair via tuple generative process.

3. **Global cohesiveness module** learns entity and relation embeddings $\mathcal{V}$ via a translating objective to capture global structure cohesiveness $\mathcal{W}$.

4. **Update sentence-level extractions** $\mathcal{T}$ based on both local context information and global structure cohesiveness.

12.2.1 THE JOINT OPTIMIZATION PROBLEM

We now show how three modules introduced above can be organically integrated. Phrase Extraction Module provide entity and relation seeds for Tuple Generation Module and Global Cohesiveness Module. Relation Tuple Generation Module can provide positive tuples for semantic representation learning, in return, global cohesiveness representation serve as good semantic measure during generation process.

Objective for Local Context. The following objective aims at finding semantic consistent tuples in each sentence s:

$$\max \sum_{(h,t)\in E_p^+} \sum_i \|w_i\| A_i(h,t), \tag{12.1}$$

where $\|w_i\| = S(h,l,t)$, $A_i(h,t) = \mathcal{P}(w_{[b_i,b_{i+1}]}|h,t)$. $\mathcal{A}$ and $\mathcal{W}$ are calculated via Phrase Extraction Module and Global Cohesiveness Module, respectively. In each sentence, it is a discrete problem to find most consistent tuples regarding given entity pairs and scores. Therefore dynamic programming is deployed to find optimal solution of Relation Tuple Generation Module.

Objective for Global Cohesiveness. With global measuring of relation tuples, we have global objective to associate extracted relation tuples in the corpus $\mathcal{D}$ as below:

$$\max \sum_{w_i\in E_{h,l,t}^+} \sum_{\tilde{w}_i\in E_{h',l,t'}^-} \|w_i\| - \|\tilde{w}_i\| - \gamma, \tag{12.2}$$

where $E_{h,l,t}^+$ denote for $(h,t)\in E_p^+$ and predicate l stands for average extracted predicate $l = (r_1, r_2, ..., r_n)$ in between, γ is the hyper margin, $E_{h',l,t'}^-$ is composed of training tuples with either **h** or **t** replaced. Global objective tries to maximize margin between positive extractions similarities and negative one's similarity, which start with current positive extractions and iteratively propagate to more unknown tuples in local optimization.

12.3 SUMMARY

This chapter studies the task of open information extraction and proposes a principled framework, ReMine, to unify local contextual information and global structural cohesiveness for effective extraction of relation tuples. ReMine leverages distant supervision in conjunction with existing KBs to provide automatically labeled sentence and guide the entity and relation segmentation. The local objective is further learned together with a translating-based objective to enforce structural cohesiveness, such that corpus-level statistics are incorporated for boosting high-quality tuples extracted from individual sentences. We develop a joint optimization algorithm to efficiently solve the proposed unified objective function and can output quality extractions by taking into account both local and global information. Experiments on two real-world corpora of different domains demonstrate that ReMine system achieves superior precision when outputting same number of extractions, compared with several state-of-the-art open IE systems.

As a byproduct, ReMine also demonstrates competitive performance on detecting mentions of entities from text when compared to several named entity recognition algorithms.

CHAPTER 13

Applications

The impact of the effort-light StructMine approach is best shown in multiple downstream applications. In this chapter we start with a discussion on how to build on top of distant supervision to incorporate human supervision (e.g., curated rules from domain experts) in the effort-light StructMine framework, followed by showing an application on life sciences domain that makes use of the StructNet constructed by our methods, and proposing a few potential new applications.

13.1 STRUCTURING LIFE SCIENCE PAPERS: THE LIFE-INET SYSTEM

Biomedical literature is one of the major sources storing biomedical knowledge and new research findings. A lot of useful information, e.g., new drugs discovered and new bio-molecular interactions, are deeply buried in the literature. Currently, the most common way to dig out such information is by human curation. For example, bio-curators will manually read each paper and try to assign the most appropriate MeSH terms for each paper to facilitate further literature retrieval. Also, they will manually extract the major biomedical entities (e.g., genes, proteins, drugs, and diseases) and their related information from each paper and add the extracted information into human curated databases (e.g., MeSH, UniProt, DrugBank, KEGG, and GO) to facilitate further biomedical research. National Institute of Health invites a group of human annotators performing manual literature annotation. This process is costly, slow, and paled by the rapid growth of biomedical literature. Developing an accurate and efficient way to automatically extract information from bio-literature may greatly facilitate biomedical research. For example, in the following sentence (taken from a PubMed publication with the PMID 383855) one can identify several biomedical entities and their relationships.

Example 13.1 Biomedical Entity Relationships These murine models demonstrate that amikacin has in vivo activity against Nocardia and may be potentially useful in the treatment of human disease.

The above sentence presents a fact that "*amikacin*" is a `chemical` entity, and claims the finding that "*amikacin*" can potentially treat "*Nocardia*," which is a `disease`. Without tools for mining entity and relation structures such as effort-light StructMine, human experts have to read through the whole sentence to identify the chemical and disease entities in the sentence, and then infer their relationship as a `treatment` relationship from the sentence. However, text mining tools, such as CoType [Ren et al., 2017a], will be able to take the large document collection and some existing biomedical databases as input, and automatically recognize "*amikacin*" as a chemical and "*Nocardia*" as a disease and further infer that there is a "treatment" relation between them. This example shows that automatic techniques for mining entity and relation structures can greatly save time, human effort and costs for biomedical information extraction from literatures, which serves as a primary step for many downstream applications such as new drug discovery, adverse event detection for drug combination, and biomedical KB construction.

As a follow-up effort, we develop a novel system, called **Life-iNet** [Ren et al., 2017c] on top of our entity recognition and relation extraction methods, which automatically turns an *unstructured* background corpus into a *structured* network of factual knowledge (see Fig. 13.1), and supports multiple exploratory and analytic functions over the constructed network for knowledge discovery. To extract factual structures, Life-iNet automatically detects token spans of entities mentioned from text (i.e., ClusType [Ren et al., 2015]), labels entity mentions with semantic categories (i.e., PLE [Ren et al., 2016a]), and identifies relationships of various relation types between the detected entities (i.e., CoType [Ren et al., 2017a]). These inter-related pieces of information are integrated to form a unified, structured network, where nodes represent different types of entities and edges denote relationships of different relation types between the entities. To address the issue of limited diversity and coverage, Life-iNet relies on the external KBs to provide seed examples (i.e., *distant supervision*), and identifies additional entities and relationships from the given corpus (e.g., using multiple textual resources such as scientific literature and encyclopedia articles) to construct a structured network. By doing so, we integrate the factual information in the existing KBs with those extracted from the given corpus. To support analytic functionality, the Life-iNet system implements link prediction functions over the construct network and integrates a distinctive summarization function to provide insight analysis (e.g., answering questions such as "*which genes are distinctively related to the given disease type under `GeneDiseaseAssociation` relation?*").

To systematically incorporate these ideas, Life-iNet leverages the novel entity and relation structure mining techniques [Ren et al., 2015, 2016a, 2017a] developed in effort-light Struct-Mine to implement an *effort-light network construction framework*. Specially, it relies on distant supervision in conjunction with external KBs to (1) detect quality entity mentions [Ren et al., 2015], (2) label entity mentions with fine-grained entity types in a given type hierarchy [Ren et al., 2016a], and (3) identify relationships of different types between entities [Ren et al., 2017a]. In particular, we design specialized loss functions to faithfully model "*appropriate*" labels and remove "*false positive*" labels for the training instances (heuristically generated by distant supervision), regarding the specific context where an instance is mentioned [Ren et al., 2016a, 2017a]. By doing so, we can construct *corpus-specific* information extraction models by using distant supervision in a noise-robust way (see Fig. 13.1). The proposed network construction framework is domain-independent—it can be quickly ported to other disciplines and sciences without ad-

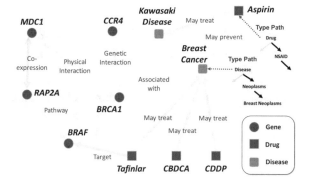

Figure 13.1: An illustrative example of the constructed Life-iNet, and its statistics.

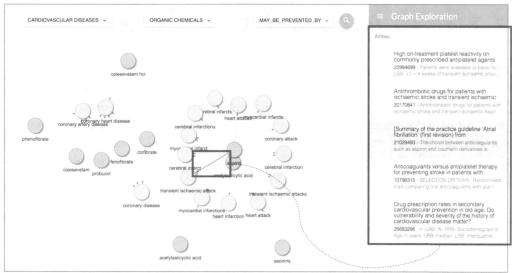

Figure 13.2: A screen-shot of the graph exploration interface of Life-iNet system. By specifying the types of two entity arguments and the relation type between them, Life-iNet system returns a graph which visualize the typed entities and relationship, and allows users to explore the graph to find relevant research papers.

ditional human labeling effort. With the constructed network, Life-iNet further applies link prediction algorithms [Bordes et al., 2013, Tang et al., 2015] to infer new entity relationships, and distinctive summarization algorithm [Tao et al., 2016] to find other entities that are distinctively related to the query entity (or the given entity types).

Impact of Life-iNet:

- A biomedical knowledge graph constructed by our Life-iNet system is used by researchers at Stanford Medical school to facilitate drug re-purposing. It yields significant improvement of performance on new drugs and rare diseases.

- Life-iNet system is adopted by veterinarians at Veterinary Information Network Inc. (VIN) to construct the first veterinary knowledge graph from multiple sources of information including research articles, books, guidelines, drug handbooks, and message board posts.

- Technologies developed in Life-iNet system have been transferred to Mayo Clinic, UCLA Medical School, and NIH Big Data to Knowledge Center to facilitate construction of domain KBs from massive scientific literature.

13.2   EXTRACTING DOCUMENT FACETS FROM TECHNICAL CORPORA

With the ever-increasing number of technical documents being generated every day, including, but not limited to, patent folios, legal cases, real-estate agreements, historical archives, and scientific literature, there is a crucial need to develop automation that can identify the *concepts* for *key facets* for each document, so that readers can quickly get a sense for what the document is about, or search and retrieve documents based on these facets. Consider the domain of scientific publications, one we are all intimately familiar with. Given a new scientific paper, it is impossible for a reader to instantly understand the *techniques* being used, the kinds of *applications* that are addressed, or the *metrics* that are used to ascertain whether the techniques have good performance. Thus, we pose the following question: *Can we develop algorithms that can efficiently and automatically identify the key facets of each document in a large technical document corpora, with little manual supervision?*

Therefore, we identify a novel research problem, called Facet Extraction, in making sense of a large corpus of technical documents. Given a collection of technical documents, the goal of facet extraction is to automatically label each document with a set of concepts for the key facets (e.g., application, technique, evaluation metrics, and dataset) that people may be interested in. The result of Facet Extraction is a summary of the major information of each document into a structured, multi-dimensional representation format, where the target facets serve as different attributes, and extracted concepts correspond to the attribute values (see Table 13.1).

Extracted facets largely enrich the original structured bibliographic meta information (e.g., authors, venues, keywords), and thus enables a wide range of interesting applications. For example, in a literature search, facets can be used to answer questions such as "which techniques are used in this paper?" and "what are the applications of this work?" (see Table 13.1), which require a deeper understanding of the paper semantics than analyzing the author-generated keyword list. One can also answer questions like "what are the popular applications in the Natural

Table 13.1: Example of extracted facets for a research publication

Technique	Application
Conditional random field	Document summarization
Unsupervised learning	Sequence labeling
Support vector machine	Statistical classification
Hidden markov model	
Evaluation Metric	**Dataset**
F1; Rouge-2	DUC

Language Processing or the Database Systems community?" and "how does the facet of *entity recognition* vary across different communities?", by aggregating the facets statistics across the database. Such results enable the discovery of ideas and the dynamics of a research topic or community in an effective and efficient way.

Our ClusType method [Ren et al., 2015] leverages relation phrase as the bridge to propagate type information. The proposed relation-based framework is general, and can be applied to different kinds of classification task. Therefore, we propose to extract document facets by doing type propagation on corpus-induced graphs. The major challenge in performing facet extraction arises from multiple sources: concept extraction, concept to facet matching, and facet disambiguation. To tackle these challenges, we extend ClusType approach and develop FacetGist, a framework for facet extraction. Facet extraction involves constructing a graph-based heterogeneous network to capture information available across multiple *local* sentence-level features, as well as *global* context features. We then formulate a joint optimization problem, and propose an efficient algorithm for graph-based label propagation to estimate the facet of each concept mention. Experimental results on technical corpora from two domains demonstrate that FacetGist can lead to an improvement of over 25% in both precision and recall over competing schemes [Siddiqui et al., 2016].

13.3   COMPARATIVE DOCUMENT ANALYSIS

In many use cases, people want to have a concise yet informative summary to describe the common and different places between two documents or two set of documents. One of our recent work presents a novel research problem, *Comparative Document Analysis* (*CDA*), that is, *joint* discovery of commonalities and differences between two individual documents (or two sets of documents) in a large text corpus. Given any pair of documents from a (background) document collection, CDA aims to automatically identify sets of quality phrases (entities) to summarize the commonalities of *both* documents and highlight the distinctions of each *with respect to the other* informatively and concisely. It makes use of the output from our entity recognition and typing method to generate candidate phrases for each document.

While there has been some research in comparative text mining, most of these focus on generating word-based or sentence-based summarization for sets of documents. Word-based summarization [Mani and Bloedorn, 1997, Zhai et al., 2004] suffers from limited readability as single words are usually non-informative and bag-of-words representation does not capture the semantics of the original document well—it may not be easy for users to interpret the combined meaning of the words. Sentence-based summarization [Huang et al., 2014, Lerman and Mc-Donald, 2009, Wang et al., 2013], on the other hand, may be too verbose to accurately highlight the *general* commonalities and differences—users may be distracted by the irrelevant information contained there (as later shown in our case study). Furthermore, previous work compares two sets of documents using redundant contents (e.g., word overlap) but the task becomes much more challenging when comparing two *individual* documents, as there exist a limited number of common content units between the two documents.

We study a novel comparative text mining problem which leverages *multi-word noun phrases* (i.e., proper names) to represent the common and distinct information between *two individual documents* (or two sets of documents), by referring to a massive background corpus for measuring semantic relevance between documents and phrases. We refer the task as Comparative Document Analysis (CDA): Given a pair of documents from a document collection, the task is to (1) extract from each document salient phrases and phrase pairs which cover its major content; (2) discover the commonalities between the document pair by selecting salient phrases which are semantically relevant to *both* of them; and (3) find the distinctions for each document by selecting salient phrases that are *exclusively* relevant to the document. CDA can benefit a variety of applications including related item recommendation and document retrieval. For example, as shown in Fig. 13.3, a citation recommendation system can show users the common and distinct concepts produced by CDA to help them understand the connections and differences between a query paper [Jeh and Widom, 2003] and a recommended paper [Haveliwala, 2003]. In a similar way, CDA can reduce the efforts on patentbility searching [Zhang et al., 2015].

Our solution uses a general graph-based framework to derive novel measures on phrase *semantic commonality* and *pairwise distinction*, where the background corpus is used for computing phrase-document semantic relevance. We use the measures to guide the selection of sets of phrases by solving two joint optimization problems. A scalable iterative algorithm is developed to integrate the maximization of phrase commonality or distinction measure with the learning of phrase-document semantic relevance. Experiments on large text corpora from two different domains—scientific papers and news—demonstrate the effectiveness and robustness of the proposed framework on comparing documents. Analysis on a 10 GB+ text corpus demonstrates the scalability of our method, whose computation time grows linearly as the corpus size increases.

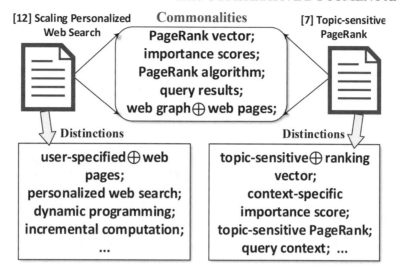

Figure 13.3: Example output of comparative document analysis (CDA) for papers [Jeh and Widom, 2003] and [Haveliwala, 2003]. CDA combines two proper names which frequently co-occur in the documents into a *name pair* using the symbol "⊕."

Our case study on comparing news articles published at different dates shows the power of the proposed method on comparing sets of documents.

CHAPTER 14

Conclusions

Entities and relationships are important structures that can be extracted from a text corpus to represent the factual knowledge inside the corpus. Effective and efficient mining of entity and relation structures from text helps gaining insights from large volume of text data (that are infeasible for human to read through and digest), and enables many downstream applications on understanding, exploring and analyzing the text content. Data analysts and government agents may want to identify person, organization and location entities in news everyday news articles and generate concise and timely summary of news events. Biomedical researchers who cannot digest large amounts of newly-published research papers in relevant areas would need an effective way to extract different relationships between proteins, drugs, and diseases so as to follow the new claims and facts presented in the research community. However, text data is highly variable: corpora covering topics from different domains, written in different genres or languages have typically required for effective processing a wide range of language resources such as grammars, vocabularies, gazetteers. The *massive* and *messy* nature of text data post significant challenges to creating tools for automated structuring of unstructured content that scale with text volume.

14.1 EFFORT-LIGHT STRUCTMINE: SUMMARY

In this book, we focus on principled and scalable methods for the mining of typed *entity* and *relation* structures from unstructured text corpora in order to overcome the barriers in dealing with text corpora of various domains, genres and languages. As traditional information extraction approaches have relied on large amounts of task-specific labeled data, my book work harnesses the power of "big text data" and focuses on creating generic solutions for *efficient construction of customized machine-learning models for factual structure extraction*, relying on only limited amounts of (or even no) task-specific training data. Our proposed methods aim to bridge the gap between customized machine-learning models and the absence of high-quality task-specific training data. It leverages the information overlap between background facts stored in external KBs (KBs) and the given corpus to automatically generate large amounts of (possibly noisy) task-specific training data; and it exploits redundant text information within the massive corpus to reduce the complexity of feature generation (e.g., sentence parsing). This solution is based on two key intuitions which are described below. Overall, the book has made the contributions on mining entity and relation structures in the following aspects.

1. We propose **three key principles** on systematically mining typed entities and relationships from massive corpora, using distant supervision in conjunction with KBs.

- **Automatic labeled data generation by aligning corpus with KBs**. In a massive corpus, structured information about some of the entities (e.g., entity types, relationships to other entities) can be found in external KBs. Can we align the corpus with external KBs to automatically generate training data for extracting entity and relation structures at a large scale? Such retrieved information supports the automated annotation of entities and relations in text and labeling of their categories, yielding (possibly noisy) corpus-specific training data. Although the overlaps between external KBs and the corpus at hand might involve only a small proportion of the corpus, the scale of the automatically labeled training data could still be much larger than that of manually annotated data by domain experts.

- **Type propagation via co-occurring text features**. Text units (e.g., word, phrase) co-occur frequently with entities and other text units in a massive corpus. We can exploit the textual co-occurrence patterns to characterize the semantics of text units, entities, and entity relations. As such patterns become more apparent in a massive corpus with rich data redundancy, big text data leads to big opportunities in representing semantics of text unit without complex feature generation. This is a principle that go through all the chapters, mainly illustrated in Chapter 4.

- **Model semantic similarity by exploiting co-occurrence patterns**. Text units used as features in type propagation framework are highly variable—one string can have multiple semantic meanings and one object can be expressed using different strings. We propose to learn the low-dimensional representations to model the semantic meaning of text units based on their surrounding context (i.e., distributional assumption). With effective semantic representation, we are able to group similar text units together to facilitate type propagation (i.e., overcome sparsity issue for infrequent text units). This is a principle that go through all the chapters, mainly illustrated in Chapters 4 and 7.

2. We study **different structure extraction tasks** for mining typed entity and relation structures from corpora, which include entity recognition and typing (Chapter 4), fine-grained entity typing (Chapter 5), and entity relationship extraction (Chapter 7). In particular, we investigate human effort-light solutions for these several tasks using distant supervision in conjunction with external KBs. This yields different problem settings as compared to the fully-supervised learning problem setup in most existing studies on information extraction. A key challenge in dealing with distant supervision is on designing effective typing models that are robust to the noisy labels in the automatically generated training data.

3. We have proposed **models** and **algorithms** to solve the above tasks.

- We studied distantly-supervised entity recognition and typing, and proposed a novel relation phrase-based entity recognition framework, ClusType (Chapter 4). A domain-agnostic phrase mining algorithm is developed for generating candidate entity mentions and relation phrases. By integrating relation phrase clustering with type propagation, the proposed method is effective in minimizing name ambiguity and context problems, and thus predicts each mention's type based on type distribution of its string name and type signatures of its surrounding relation phrases. We formulate a joint optimization problem to learn object type indicators/signatures and cluster memberships simultaneously.

- For fine-grained entity typing, we propose hierarchical partial-label embedding methods, AFET and PLE , that models "clean" and "noisy" mentions separately and incorporates a given type hierarchy to induce loss functions (Chapter 5). Both models build on a joint optimization framework, learns embeddings for mentions and type-paths, and iteratively refines the model.

- Our work on extracting typed relationships studies domain-independent, joint extraction of typed entities and relationships with distant supervision (Chapter 7). The proposed Co-Type framework runs domain-agnostic segmentation algorithm to mine entity mentions, and formulates the joint entity and relation mention typing problem as a global embedding problem. We design a noise-robust objective to faithfully model noisy type label from distant supervision, and capture the mutual dependencies between entity and relation based on the translation embedding assumption.

14.2 CONCLUSION

The contributions of this book work are in the area of text mining and information extraction, within which we focus on domain-independent and noise-robust approaches using distant supervision (in conjunction with publicly-available KBs). The work has broad impact on a variety of applications: KB construction, question-answering systems, structured search and exploration of text data, recommender systems, network analysis, and many other text mining tasks. Finally, our work has been used in the following settings.

- **Introduced in classes and conference tutorials**: Our methods on entity recognition and typing (ClusType), fine-grained entity typing (PLE [Ren et al., 2016a], AFET [Ren et al., 2016b]), and relation extraction (CoType [Ren et al., 2017a]) are being taught in graduate courses, e.g., University of Illinois at Urbana-Champaign (CS 512), and are introduced as major parts of the conference tutorial in top data mining and database conferences such as SIGKDD, WWW, CIKM, and SIGMOD.

- **Real-world, cross-disciplinary use cases**:

  – Our entity recognition and typing technique (ClusType [Ren et al., 2014]) has been transferred to U.S. Army Research Lab, Microsoft Bing Ads and NIH Big Data to Knowledge Center to identify typed entities of different kinds from low-resource, domain-specific text corpora. ClusType is also used by Stanford sociologists to identify scientific concepts from 37 millions of scientific publications in Web of Science database to study innovation and translation of scientific ideas.

    – A biomedical knowledge graph (i.e., Life-iNet [Ren et al., 2017c]) constructed automatically from millions of PubMed publications using our effort-light StructMine pipeline is used by researchers at Stanford Medical school to facilitate drug re-purposing. It yields significant improvement of performance on new drugs and rare diseases.

    – Our effort-light StructMine techniques (ClusType, PLE, CoType) is adopted by veterinarians at Veterinary Information Network Inc. (VIN) to construct the first veterinary knowledge graph from multiple sources of information including research articles, books, guidelines, drug handbooks, and message board posts.

- **Awards**: The book work on effort-light StructMine has been awarded a Google Ph.D. fellowship in 2016 (sole winner in the category of Structured Data and Data Management in the world) and a Yahoo!-DAIS Research Excellence Award, and a C. W. Gear Outstanding Graduate Student Award from University of Illinois.

CHAPTER 15

Vision and Future Work

This chapter discusses several potential directions for future work. Along this line of research, there are three exciting directions that could be pursued: (1) exploring more ways to reduce human annotation efforts other than leverage distant supervision from KBs; (2) extract implicit language patterns from massive, unlabeled corpora to facilitate supervised models; and (3) enrich the factual structures currently defined for the corpus-specific StructNet to enable more use cases.

15.1 EXTRACTING IMPLICIT PATTERNS FROM MASSIVE UNLABELED CORPORA

As neural language models can be trained without human annotations but generate texts of a high quality, we further explore the possibility to extract the abundant and self-contained knowledge in natural language. So far, we've employed two strategies to incorporate such models with sequence labeling, a general framework in natural language processing which encompassing various of applications (e.g., Named Entity Recognition, POS Tagging, Event Detection). The first is to treat such information as additional supervision, and guide the training of the target task by the knowledge transfer.

Specifically, we leave the word-level knowledge to pre-trained word embedding and co-train a character-level language model with the end task. The proposed method can conduct efficient training and inference, which has been accepted and presented at the AAAI 2017 conference. Alternatively, we further explore the potential of the extensive raw corpora. We pre-train very large language models on such corpora to capture abundant linguistic features. Moreover, we design a novel model pruning method to allow us conduct model compression for better inference efficiency. The resulting model can be incorporated in a plug-in-and-play manner and greatly improve the performance without much loss of efficiency. This work has been submitted to a reputed venue for the review.

15.2 ENRICHING FACTUAL STRUCTURE REPRESENTATION

In the current definition of StructNet, edges between two entities are weighted by the frequency of the facts mentioned in the text corpus. Such a representation has several limitations: (1) raw frequency cannot indicate *uncertainty* of the fact (e.g., drug A treats drug B with 75% success rate), (2) conditions of a relation are ignored in the modeling (e.g., if the patient is under 50 years old), and (3) complex relations involve more than two entities (e.g., protein `localization` relation). To address these challenges, I am interested in **collaborating with NLP researchers and linguists** to work on domain-independent sentiment analysis and syntax parsing for large text corpora, and incorporate the sophisticated linguistic features in StructNet construction. In particular, to measure fact uncertainty, it is critical to mine from a sentence words/phrases that indicate uncertainty (e.g., "*unlikely*," "*probably*," "*with 50% chance*"), negation (e.g., "*no*," "*barely*"), sentiments (e.g., "*efficiently*," "*nicely*"), or their enhancers (e.g., "very," "extremely"), and design systematic measures to quantify these units into weights of the edges in StructNets. To mine conditions for relations, I aim to extend the meta pattern-based attribute mining algorithm to identify patterns for "condition descriptions" (e.g., "...*[with age _]*...") and attach the mined conditions to edges in StructNet for further analysis. To extract complex relations, I plan to design scalable candidate generation process (e.g., different pruning strategy) to avoid producing exponential number of candidate relations, and extend the CoType embedding approach to model types for n-ary relations, while preserving the mutual constraints between relations and their entity arguments.

Bibliography

E. Agichtein and L. Gravano (2000. Snowball: Extracting relations from large plain-text collections, in *Proc. of the 5th ACM Conference on Digital Libraries*. DOI: 10.1145/336597.336644. 22, 29, 112

R. Agrawal and R. Srikant (1995). Mining sequential patterns, in *ICDE*, pp. 3–14. DOI: 10.1109/icde.1995.380415. 141, 142

G. Angeli, M. J. J. Premkumar, and C. D. Manning (2015). Leveraging linguistic structure for open domain information extraction, in *Proc. of the 53rd Annual Meeting of the Association for Computational Linguistics and the 7th International Joint Conference on Natural Language Processing (Volume 1: Long Papers)*, vol. 1, pp. 344–354. DOI: 10.3115/v1/p15-1034. 31, 148

R. Angheluta, R. De Busser, and M.-F. Moens (2002). The use of topic segmentation for automatic summarization, in *Proc. of the ACL Workshop on Automatic Summarization*, pp. 11–12. 75

D. E. Appelt, J. R. Hobbs, J. Bear, D. Israel, and M. Tyson (1992). Fastus: A finite-state processor for information extraction from real-world text, in *IJCAI*.

Y. Artzi and L. S. Zettlemoyer (2013). Weakly supervised learning of semantic parsers for mapping instructions to actions, *TACL*, vol. 1, pp. 49–62. 136

S. Auer, C. Bizer, G. Kobilarov, J. Lehmann, R. Cyganiak, and Z. Ives (2007). Dbpedia: A nucleus for a web of open data, in *The Semantic Web*, pp. 722–735, Springer. DOI: 10.1007/978-3-540-76298-0_52. 22, 139

N. Bach and S. Badaskar (2007). A review of relation extraction, *Literature Review for Language and Statistics II*. 28, 87, 105

M. Banko, M. J. Cafarella, S. Soderland, M. Broadhead, and O. Etzioni (2007). Open information extraction from the web, in *IJCAI*. 3, 22, 31, 139, 147, 148

J. Bao, N. Duan, M. Zhou, and T. Zhao (2014). Knowledge-based question answering as machine translation, *Cell*, vol. 2, no. 6. DOI: 10.3115/v1/p14-1091. 120

J. Bian, Y. Liu, E. Agichtein, and H. Zha (2008). Finding the right facts in the crowd: Factoid question answering over social media, in *WWW*. DOI: 10.1145/1367497.1367561. 87

L. Bing, S. Chaudhari, R. Wang, and W. Cohen (2015). Improving distant supervision for information extraction using label propagation through lists, in *Proc. of the Conference on Empirical Methods in Natural Language Processing*, pp. 524–529. DOI: 10.18653/v1/d15-1060. 112

A. Blum and T. Mitchell (1998). Combining labeled and unlabeled data with co-training, in *COLT Workshop on Computational Learning Theory*. DOI: 10.1145/279943.279962. 29, 113

K. Bollacker, C. Evans, P. Paritosh, T. Sturge, and J. Taylor (2008). Freebase: A collaboratively created graph database for structuring human knowledge, in *SIGMOD*. DOI: 10.1145/1376616.1376746. 4, 22, 35, 38, 90, 139

A. Bordes, N. Usunier, A. Garcia-Duran, J. Weston, and O. Yakhnenko (2013). Translating embeddings for modeling multi-relational data, in *NIPS*. 30, 92, 96, 100, 149, 155

L. Breiman (2001). Random forests, *Machine Learning*, vol. 45, no. 1, pp. 5–32. DOI: 10.1023/A:1010933404324. 143

S. Brin (1998). Extracting patterns and relations from the world wide web, in *International Workshop on the World Wide Web and Databases*. DOI: 10.1007/10704656_11. 29

R. C. Bunescu and R. J. Mooney (2005). A shortest path dependency kernel for relation extraction, in *HLT-EMNLP*. DOI: 10.3115/1220575.1220666.

R. C. Bunescu and R. Mooney (2007). Learning to extract relations from the web using minimal supervision, in *ACL*. 87

R. C. Bunescu and R. J. Mooney (2007). Learning to extract relations from the Web using minimal supervision, in *ACL*.

A. Carlson, J. Betteridge, B. Kisiel, B. Settles, E. R. Hruschka Jr., and T. M. Mitchell (2010). Toward an architecture for never-ending language learning, in *AAAI*. 2, 3, 22, 139

A. Carlson, J. Betteridge, R. C. Wang, E. R. Hruschka Jr., and T. M. Mitchell (2010). Coupled semi-supervised learning for information extraction, in *WSDM*. DOI: 10.1145/1718487.1718501. 119, 147, 148

Y. S. Chan and D. Roth (2010). Exploiting background knowledge for relation extraction, in *COLING*. 94, 133

Z. Chen, A. Mukherjee, and B. Liu (2014). Aspect extraction with automated prior knowledge learning, in *ACL*. DOI: 10.3115/v1/p14-1033. 139

L. Del Corro and R. Gemulla (2013). Clausie: Clause-based open information extraction, in *WWW*. DOI: 10.1145/2488388.2488420. 148, 150

T. Cour, B. Sapp, and B. Taskar (2011). Learning from partial labels, *JMLR*, vol. 12, pp. 1501–1536. 30, 70

A. Culotta and J. Sorensen (2004). Dependency tree kernels for relation extraction, in *ACL*. DOI: 10.3115/1218955.1219009. 28, 87, 111

J. R. Curran, T. Murphy, and B. Scholz (2007). Minimising semantic drift with mutual exclusion bootstrapping, in *PACLING*, pp. 172–180. 112, 116

G. R. Doddington, A. Mitchell, M. A. Przybocki, L. A. Ramshaw, S. Strassel, and R. M. Weischedel (2004). The automatic content extraction (ace) program-tasks, data, and evaluation, in *LREC*, vol. 2, p. 1. 20, 21

X. L. Dong, T. Strohmann, S. Sun, and W. Zhang (2014). Knowledge vault: A web-scale approach to probabilistic knowledge fusion, in *SIGKDD*. DOI: 10.1145/2623330.2623623. 2, 35, 59, 87

L. Dong, F. Wei, H. Sun, M. Zhou, and K. Xu (2015). A hybrid neural model for type classification of entity mentions, in *IJCAI*. 70

H. Drucker, C. J. Burges, L. Kaufman, A. Smola, V. Vapnik et al. (1997). Support vector regression machines, *NIPS*, vol. 9, pp. 155–161. 145

J. Dunietz and D. Gillick (2014). A new entity salience task with millions of training examples, *EACL*. DOI: 10.3115/v1/e14-4040. 61

A. El-Kishky, Y. Song, C. Wang, C. R. Voss, and J. Han (2014). Scalable topical phrase mining from text corpora, *VLDB*, vol. 8, no. 3, pp. 305–316. DOI: 10.14778/2735508.2735519. 93

A. El-Kishky, Y. Song, C. Wang, C. R. Voss, and J. Han (2015). Scalable topical phrase mining from text corpora, *VLDB*. DOI: 10.14778/2735508.2735519. 40, 41

J. Ellis, J. Getman, J. Mott, X. Li, K. Griffitt, S. M. Strassel, and J. Wright (2014). Linguistic resources for 2013 knowledge base population evaluations, *Text Analysis Conference (TAC)*. 103, 134

O. Etzioni, M. Cafarella, D. Downey, S. Kok, A.-M. Popescu, T. Shaked, S. Soderland, D. S. Weld, and A. Yates (2004). Web-scale information extraction in knowitall (preliminary results), in *WWW*. DOI: 10.1145/988672.988687. 2, 3, 87

O. Etzioni, M. Cafarella, D. Downey, A.-M. Popescu, T. Shaked, S. Soderland, D. S. Weld, and A. Yates (2005). Unsupervised named-entity extraction from the Web: An experimental study, *Artificial Intelligence*, vol. 165, no. 1, pp. 91–134. DOI: 10.1016/j.artint.2005.03.001.

A. Fader, S. Soderland, and O. Etzioni (2011). Identifying relations for open information extraction, in *EMNLP*. 31, 38, 40, 53, 139, 149

J. R. Finkel, T. Grenager, and C. Manning (2005). Incorporating non-local information into information extraction systems by Gibbs sampling, in *ACL*. DOI: 10.3115/1219840.1219885. 52, 62, 93

E. Gabrilovich, M. Ringgaard, and A. Subramanya (2013). Facc1: Freebase annotation of clueweb corpora. 62

L. Galárraga, G. Heitz, K. Murphy, and F. M. Suchanek (2014). Canonicalizing open knowledge bases, in *CIKM*. DOI: 10.1145/2661829.2662073. 31, 46

R. Ghani, K. Probst, Y. Liu, M. Krema, and A. Fano (2006). Text mining for product attribute extraction, *SIGKDD Explorations*, vol. 8, no. 1, pp. 41–48. DOI: 10.1145/1147234.1147241. 139

D. Gillick, N. Lazic, K. Ganchev, J. Kirchner, and D. Huynh (2014). Context-dependent fine-grained entity type tagging, *ArXiv Preprint ArXiv:1412.1820*. 60, 61, 69, 70, 91, 104

M. R. Gormley, M. Yu, and M. Dredze (2015). Improved relation extraction with feature-rich compositional embedding models, *EMNLP*. DOI: 10.18653/v1/d15-1205. 30, 92, 104

Z. GuoDong, S. Jian, Z. Jie, and Z. Min (2005). Exploring various knowledge in relation extraction, in *ACL*. DOI: 10.3115/1219840.1219893. 28, 87, 92, 95

S. Gupta and C. D. Manning (2014). Improved pattern learning for bootstrapped entity extraction, in *CONLL*. DOI: 10.3115/v1/w14-1611. 29, 35, 52

R. Gupta, A. Halevy, X. Wang, S. E. Whang, and F. Wu (2014). Biperpedia: An ontology for search applications, *Proc. of the VLDB Endowment*, vol. 7, no. 7, pp. 505–516. DOI: 10.14778/2732286.2732288. 139

C. Van Gysel, M. de Rijke, and E. Kanoulas (2016a). Learning latent vector spaces for product search, in *Proc. of the 25th ACM International on Conference on Information and Knowledge Management*, pp. 165–174, ACM. DOI: 10.1145/2983323.2983702. 124

C. Van Gysel, M. de Rijke, and M. Worring (2016b). Unsupervised, efficient and semantic expertise retrieval, in *Proc. of the 25th International Conference on World Wide Web*, pp. 1069–1079, International World Wide Web Conferences Steering Committee. DOI: 10.1145/2872427.2882974. 124

F. Harary and I. C. Ross (1957). A procedure for clique detection using the group matrix, *Sociometry*, vol. 20, no. 3, pp. 205–215. DOI: 10.2307/2785673. 144

Z. S. Harris (1954). Distributional structure, *Word*, vol. 10, pp. 146–162. DOI: 10.1080/00437956.1954.11659520. 122

T. H. Haveliwala (2003). Topic-sensitive pagerank: A context-sensitive ranking algorithm for web search, *TKDE*, vol. 15, no. 4, pp. 784–796. DOI: 10.1109/tkde.2003.1208999. 158, 159

X. He and P. Niyogi (2004). Locality preserving projections, in *NIPS*. 44, 47, 96, 97

Y. He and D. Xin (2011). Seisa: Set expansion by iterative similarity aggregation, in *WWW*. DOI: 10.1145/1963405.1963467.

M. A. Hearst (1992). Automatic acquisition of hyponyms from large text corpora, in *Proc. of the 14th Conference on Computational Linguistics (ACL)*. DOI: 10.3115/992133.992154. 77, 139, 149

J. Hoffart, M. A. Yosef, I. Bordino, H. Fürstenau, M. Pinkal, M. Spaniol, B. Taneva, S. Thater, and G. Weikum (2011). Robust disambiguation of named entities in text, in *EMNLP*. 91

R. Hoffmann, C. Zhang, X. Ling, L. Zettlemoyer, and D. S. Weld (2011). Knowledge-based weak supervision for information extraction of overlapping relations, in *ACL*. 87, 94, 102, 104, 105

R. Hoffmann, C. Zhang, X. Ling, L. Zettlemoyer, and D. S. Weld (2011). Knowledge-based weak supervision for information extraction of overlapping relations, in *ACL*, pp. 541–550. 112

R. Hoffmann, C. Zhang, X. Ling, L. S. Zettlemoyer, and D. S. Weld (2011). Knowledge-based weak supervision for information extraction of overlapping relations, in *ACL*. 130, 134

D. Hovy, B. Plank, H. M. Alonso, and A. Søgaard (2015). Mining for unambiguous instances to adapt part-of-speech taggers to new domains, in *NAACL*. DOI: 10.3115/v1/n15-1135. 89

M. Hu and B. Liu (2014). Mining and summarizing customer reviews, in *KDD*. DOI: 10.1145/1014052.1014073. 139

R. Huang and E. Riloff (2010). Inducing domain-specific semantic class taggers from (almost) nothing, in *ACL*. 35, 36, 52

X. Huang, X. Wan, and J. Xiao (2014). Comparative news summarization using concept-based optimization, *Knowledge and Information Systems*, vol. 38, no. 3, pp. 691–716. DOI: 10.1007/s10115-012-0604-8. 158

L. Huang, J. May, X. Pan, H. Ji, X. Ren, J. Han, L. Zhao, and J. A. Hendler (2017). Liberal entity extraction: Rapid construction of fine-grained entity typing systems, *Big Data*, vol. 5, no. 1, pp. 19–31. DOI: 10.1089/big.2017.0012.

G. Jeh and J. Widom (2003). Scaling personalized web search, in *WWW*. DOI: 10.1145/775189.775191. 158, 159

H. Ji and R. Grishman (2008). Refining event extraction through cross-document inference, in *ACL*. 59

J.-Y. Jiang, C.-Y. Lin, and P.-J. Cheng (2015). Entity-driven type hierarchy construction for freebase, in *WWW*. DOI: 10.1145/2740908.2742737. 65

A. Kannan, I. E. Givoni, R. Agrawal, and A. Fuxman (2011). Matching unstructured product offers to structured product specifications, in *KDD*, pp. 404–412. DOI: 10.1145/2020408.2020474. 139

Z. Kozareva and E. Hovy (2010). Not all seeds are equal: Measuring the quality of text mining seeds, in *NAACL*. 35

C. Lee, Y.-G. Hwang, and M.-G. Jang (2007). Fine-grained named entity recognition and relation extraction for question answering, in *SIGIR*. DOI: 10.1145/1277741.1277915. 91

D. B. Lenat (1995). Cyc: A large-scale investment in knowledge infrastructure, *Communications of the ACM*, vol. 38, no. 11, pp. 33–38. DOI: 10.1145/219717.219745. 22

K. Lerman and R. McDonald (2009). Contrastive summarization: An experiment with consumer reviews, in *NAACL*. DOI: 10.3115/1620853.1620886. 158

Q. Li and H. Ji (2014). Incremental joint extraction of entity mentions and relations, in *ACL*. DOI: 10.3115/v1/p14-1038. 28, 87, 104

Y. Li, J. Gao, C. Meng, Q. Li, L. Su, B. Zhao, W. Fan, and J. Han (2016). A survey on truth discovery, *SIGKDD Explore Newsletter*, vol. 17, no. 2, pp. 1–16, http://doi.acm.org/10.1145/2897350.2897352 DOI: 10.1145/2897350.2897352. 119

T. Lin et al. (2012). No noun phrase left behind: Detecting and typing unlinkable entities, in *EMNLP*. 30, 31, 35, 36, 38, 53, 59, 91

Y. Lin, S. Shen, Z. Liu, H. Luan, and M. Sun (2016). Neural relation extraction with selective attention over instances. in *ACL*, pp. 2124–2133. DOI: 10.18653/v1/p16-1200. 112

Y. Lin, S. Shen, Z. Liu, H. Luan, and M. Sun (2016). Neural relation extraction with selective attention over instances, in *ACL*. DOI: 10.18653/v1/p16-1200. 130

X. Ling and D. S. Weld (2012). Fine-grained entity recognition, in *AAAI*. 29, 30, 53, 59, 60, 62, 64, 69, 70, 94, 95, 98, 102, 103, 104, 105, 109, 119, 134, 139

J. Liu, C. Wang, J. Gao, and J. Han (2013). Multi-view clustering via joint nonnegative matrix factorization, in *Proc. of the SIAM International Conference on Data Mining (SDM)*. DOI: 10.1137/1.9781611972832.28. 47, 49

J. Liu, J. Shang, C. Wang, X. Ren, and J. Han (2015). Mining quality phrases from massive text corpora, in *Proc. of the ACM SIGMOD International Conference on Management of Data (SIGMOD)*. DOI: 10.1145/2723372.2751523. 93, 94, 141

Y. Liu, F. Wei, S. Li, H. Ji, M. Zhou, and H. Wang (2015). A dependency-based neural network for relation classification, in *ACL*, pp. 285–290. DOI: 10.3115/v1/p15-2047. 112

L. Liu, X. Ren, Q. Zhu, S. Zhi, H. Gui, H. Ji, and J. Han (2017). Heterogeneous supervision for relation extraction: A representation learning approach, in *Proc. of the Conference on Empirical Methods in Natural Language Processing (EMNLP)*. DOI: 10.18653/v1/d17-1005. 13

I. Mani and E. Bloedorn (1997). Multi-document summarization by graph search and matching, *AAAI*. 158

C. D. Manning, M. Surdeanu, J. Bauer, J. Finkel, S. J. Bethard, and D. McClosky (2014). The stanford corenlp natural language processing toolkit, *ACL*. DOI: 10.3115/v1/p14-5010. 69, 103, 132

M.-C. De Marneffe, B. MacCartney, C. D. Manning et al. (2006). Generating typed dependency parses from phrase structure parses, in *Proc. of LREC*, vol. 6, pp. 449–454, Genoa. 139

R. McDonald, K. Crammer, and F. Pereira (2005). Online large-margin training of dependency parsers, in *ACL*, pp. 91–98. DOI: 10.3115/1219840.1219852. 140

P. N. Mendes, M. Jakob, A. García-Silva, and C. Bizer (2011). Dbpedia spotlight: Shedding light on the web of documents, in *I-Semantics*. DOI: 10.1145/2063518.2063519. 91, 92

T. Mikolov, I. Sutskever, K. Chen, G. S. Corrado, and J. Dean (2013). Distributed representations of words and phrases and their compositionality, in *NIPS*. 30, 77, 96, 97, 123, 145

T. Mikolov, K. Chen, G. Corrado, and J. Dean (2013). Efficient estimation of word representations in vector space, *ArXiv Preprint ArXiv:1301.3781*. 113

B. Min, S. Shi, R. Grishman, and C.-Y. Lin (2012). Ensemble semantics for large-scale unsupervised relation extraction, in *EMNLP*. 46

M. Mintz, S. Bills, R. Snow, and D. Jurafsky (2009). Distant supervision for relation extraction without labeled data, in *ACL*. DOI: 10.3115/1690219.1690287. 30, 59, 76, 87, 94, 104, 112, 119, 122

M. Mintz, S. Bills, R. Snow, and D. Jurafsky (2009). Distant supervision for relation extraction without labeled data, in *ACL/IJCNLP*. DOI: 10.3115/1690219.1690287. 130, 133

M. Miwa and Y. Sasaki (2014). Modeling joint entity and relation extraction with table representation, in *EMNLP*. DOI: 10.3115/v1/d14-1200. 28, 87

M. Miwa and M. Bansal (2016). End-to-end relation extraction using LSTMS on sequences and tree structures, *ArXiv Preprint ArXiv:1601.00770*. DOI: 10.18653/v1/p16-1105.

R. J. Mooney and R. C. Bunescu (2005). Subsequence kernels for relation extraction, in *NIPS*. 104, 105

R. J. Mooney and R. C. Bunescu (2006). Subsequence kernels for relation extraction, in *NIPS*, pp. 171–178, MIT Press. 111

D. Nadeau and S. Sekine (2007). A survey of named entity recognition and classification, *Lingvisticae Investigationes*, vol. 30, no. 1, pp. 3–26. DOI: 10.1075/bct.19.03nad. 28, 35, 59, 93, 141

N. Nakashole, M. Theobald, and G. Weikum (2011). Scalable knowledge harvesting with high precision and high recall, in *Proc. of the 4th ACM International Conference on Web Search and Data Mining*, pp. 227–236, ACM. DOI: 10.1145/1935826.1935869. 119

N. Nakashole, G. Weikum, and F. Suchanek (2012). Patty: A taxonomy of relational patterns with semantic types, in *EMNLP*. 75, 112, 139, 144, 150

N. Nakashole, T. Tylenda, and G. Weikum (2013). Fine-grained semantic typing of emerging entities, in *ACL*. 30, 35, 36, 38, 87, 139

V. Nastase, M. Strube, B. Börschinger, C. Zirn, and A. Elghafari (2010). Wikinet: A very large scale multi-lingual concept network, in *LREC*. 139

N. Nguyen and R. Caruana (2008). Classification with partial labels, in *KDD*. DOI: 10.1145/1401890.1401958. 30, 67, 70, 92, 96, 98, 133

K. Nigam and R. Ghani (2000). Analyzing the effectiveness and applicability of co-training, in *CIKM*. DOI: 10.1145/354756.354805. 29

N. F. Noy, N. H. Shah, P. L. Whetzel, B. Dai, M. Dorf, N. Griffith, C. Jonquet, D. L. Rubin, M.-A. Storey, C. G. Chute, and M. A. Musen (2009). Bioportal: Ontologies and integrated data resources at the click of a mouse, *Nucleic Acids Research*, vol. 37, pp. W170–3. DOI: 10.1093/nar/gkp440. 4

M. Pasca and B. Van Durme (2008). Weakly-supervised acquisition of open-domain classes and class attributes from web documents and query logs, in *ACL*, pp. 19–27. 139

J. Pei, J. Han, B. Mortazavi-Asl, J. Wang, H. Pinto, Q. Chen, U. Dayal, and M.-C. Hsu (2004). Mining sequential patterns by pattern-growth: The prefixspan approach, *TKDE*, vol. 16, no. 11, pp. 1424–1440. DOI: 10.1109/tkde.2004.77. 142

J. Pennington, R. Socher, and C. D. Manning (2014). Glove: Global vectors for word representation, in *EMNLP*, vol. 14, pp. 1532–43. DOI: 10.3115/v1/d14-1162. 77, 113

B. Perozzi, R. Al-Rfou, and S. Skiena (2014). Deepwalk: Online learning of social representations, in *KDD*. DOI: 10.1145/2623330.2623732. 30, 70, 104

K. Probst, R. Ghani, M. Krema, A. Fano, and Y. Liu (2007). Semi-supervised learning of attribute-value pairs from product descriptions, in *AAAI*. 139

M. Purver and S. Battersby (2012). Experimenting with distant supervision for emotion classification, in *Proc. of the 13th Conference of the European Chapter of the Association for Computational Linguistics*, pp. 482–491. 76

S. Pyysalo, F. Ginter, J. Heimonen, J. Björne, J. Boberg, J. Järvinen, and T. Salakoski (2007). Bioinfer: A corpus for information extraction in the biomedical domain, *BMC Bioinformatics*, vol. 8, no. 1, p. 1. DOI: 10.1186/1471-2105-8-50. 103

L. Qian, G. Zhou, F. Kong, and Q. Zhu (2009). Semi-supervised learning for semantic relation classification using stratified sampling strategy, in *Proc. of the Conference on Empirical Methods in Natural Language Processing: Volume 3*, pp. 1437–1445, Association for Computational Linguistics. DOI: 10.3115/1699648.1699690. 77

M. Qu, X. Ren, and J. Han (2017). Automatic synonym discovery with knowledge bases, in *Proc. of the 23rd ACM SIGKDD International Conference on Knowledge Discovery and Data Mining (KDD)*. DOI: 10.1145/3097983.3098185. 10

M. Qu, X. Ren, and J. Han (2017). Automatic synonym discovery with knowledge bases, in *KDD*, pp. 997–1005. DOI: 10.1145/3097983.3098185. 112, 113

J. Rao, H. He, and J. J. Lin (2016). Noise-contrastive estimation for answer selection with deep neural networks, in *CIKM*. DOI: 10.1145/2983323.2983872. 133

L. Ratinov and D. Roth (2009). Design challenges and misconceptions in named entity recognition, in *ACL*. DOI: 10.3115/1596374.1596399. 28, 35, 59

A. J. Ratner, C. M. De Sa, S. Wu, D. Selsam, and C. Ré (2016). Data programming: Creating large training sets, quickly, in *Advances in Neural Information Processing Systems*, pp. 3567–3575. 13, 119, 121

S. Ravi and M. Paşca (2008). Using structured text for large-scale attribute extraction, in *CIKM*, pp. 1183–1192. DOI: 10.1145/1458082.1458238. 139

X. Ren, J. Liu, X. Yu, U. Khandelwal, Q. Gu, L. Wang, and J. Han (2014). ClusCite: Effective citation recommendation by information network-based clustering, in *Proc. of the 20th ACM SIGKDD International Conference on Knowledge Discovery and Data Mining (KDD)*. DOI: 10.1145/2623330.2623630. 163

X. Ren, A. El-Kishky, C. Wang, F. Tao, C. R. Voss, and J. Han (2015). ClusType: Effective entity recognition and typing by relation phrase-based clustering, in *Proc. of the 21th ACM SIGKDD International Conference on Knowledge Discovery and Data Mining (KDD)*. DOI: 10.1145/2783258.2783362. 7, 70, 76, 91, 96, 119, 154, 157

X. Ren, W. He, M. Qu, H. Ji, C. R. Voss, and J. Han (2016a). Label noise reduction in entity typing by heterogeneous partial-label embedding, in *Proc. of the 22nd ACM SIGKDD International Conference on Knowledge Discovery and Data Mining (KDD)*. DOI: 10.1145/2939672.2939822. 8, 9, 154, 163

X. Ren, W. He, M. Qu, H. Ji, and J. Han (2016b). AFET: Automatic fine-grained entity typing by hierarchical partial-label embedding, in *Proc. of the Conference on Empirical Methods in Natural Language Processing (EMNLP)*. DOI: 10.18653/v1/d16-1144. 8, 9, 163

X. Ren, W. He, M. Qu, C. R. Voss, H. Ji, and J. Han (2016c). Label noise reduction in entity typing by heterogeneous partial-label embedding, in *KDD*. DOI: 10.1145/2939672.2939822. 30, 92, 95, 104, 105, 110

X. Ren, A. El-Kishky, C. Wang, and J. Han (2016d). Automatic entity recognition and typing in massive text corpora, in *WWW*. DOI: 10.1145/2872518.2891065. 59

X. Ren, Z. Wu, W. He, M. Qu, C. R. Voss, H. Ji, T. F. Abdelzaher, and J. Han (2016e). Cotype: Joint extraction of typed entities and relations with knowledge bases, *ArXiv Preprint ArXiv:1610.08763*. DOI: 10.1145/3038912.3052708. 122, 126

X. Ren, Z. Wu, W. He, M. Qu, C. R. Voss, H. Ji, T. F. Abdelzaher, and J. Han (2017a). CoType: Joint extraction of typed entities and relations with knowledge bases, in *Proc. of the 27th International Conference on World Wide Web (WWW)*. DOI: 10.1145/3038912.3052708. 11, 111, 154, 163

X. Ren, Z. Wu, W. He, M. Qu, C. R. Voss, H. Ji, T. F. Abdelzaher, and J. Han (2017b). Cotype: Joint extraction of typed entities and relations with knowledge bases, in *WWW*. DOI: 10.1145/3038912.3052708.

X. Ren, J. Shen, M. Qu, X. Wang, Z. Wu, Q. Zhu, M. Jiang, F. Tao, S. Sinha, D. Liem et al. (2017c). Life-inet: A structured network-based knowledge exploration and analytics system for life sciences, *Proc. of ACL System Demonstrations*. DOI: 10.18653/v1/p17-4010. 154, 164

S. Riedel, L. Yao, and A. McCallum (2010). Modeling relations and their mentions without labeled text, in *ECML*. DOI: 10.1007/978-3-642-15939-8_10. 30, 88, 95, 102, 103

S. Riedel, L. Yao, and A. McCallum (2010). Modeling relations and their mentions without labeled text, in *ECML/PKDD*. DOI: 10.1007/978-3-642-15939-8_10. 130, 134

S. Riedel, L. Yao, A. McCallum, and B. M. Marlin (2013). Relation extraction with matrix factorization and universal schemas. in *HLT-NAACL*, pp. 74–84. 112

S. Riedel, L. Yao, A. McCallum, and B. M. Marlin (2013). Relation extraction with matrix factorization and universal schemas, in *NAACL*. 136

S. Roller, K. Erk, and G. Boleda (2014). Inclusive yet selective: Supervised distributional hypernymy detection, in *COLING*, pp. 1025–1036. 75, 77

D. Roth and W.-t. Yih (2007). Global inference for entity and relation identification via a linear programming formulation, *Introduction to statistical relational learning*, pp. 553–580. 28, 87

B. Salehi, P. Cook, and T. Baldwin (2015). A word embedding approach to predicting the compositionality of multiword expressions, in *NAACL-HLT*. DOI: 10.3115/v1/n15-1099. 30

S. Sarawagi and W. W. Cohen (2004). Semi-Markov conditional random fields for information extraction, in *NIPS*. 28

S. Sarawagi (2008). Information extraction, *Foundations and Trends in Databases*, vol. 1, no. 3, pp. 261–377. DOI: 10.1561/1900000003. 147

M. Schmitz, R. Bart, S. Soderland, O. Etzioni et al. (2012). Open language learning for information extraction, in *EMNLP*. 31, 35, 112, 147, 148, 150

S. Shalev-Shwartz, Y. Singer, N. Srebro, and A. Cotter (2011). Pegasos: Primal estimated sub-gradient solver for SVM, *Mathematical Programming*, vol. 127, no. 1, pp. 3–30. DOI: 10.1007/s10107-010-0420-4. 101, 102, 134

W. Shen, J. Wang, P. Luo, and M. Wang (2012). A graph-based approach for ontology population with named entities, in *CIKM*. DOI: 10.1145/2396761.2396807. 36, 53

W. Shen, J. Wang, and J. Han (2014). Entity linking with a knowledge base: Issues, techniques, and solutions, *TKDE*, no. 99, pp. 1–20. DOI: 10.1109/tkde.2014.2327028. 35, 39

S. Shi, H. Zhang, X. Yuan, and J.-R. Wen (2010). Corpus-based semantic class mining: Distributional vs. pattern-based approaches, in *COLING*. 29

J. Shin, S. Wu, F. Wang, C. De Sa, C. Zhang, and C. Ré (2015). Incremental knowledge base construction using deepdive, *Proc. of the VLDB Endowment*, vol. 8, no. 11, pp. 1310–1321. DOI: 10.14778/2809974.2809991. 2

V. Shwartz, Y. Goldberg, and I. Dagan (2016). Improving hypernymy detection with an integrated path-based and distributional method, in *ACL*, pp. 2389–2398. DOI: 10.18653/v1/p16-1226. 112, 113

V. Shwartz and I. Dagan (2016). Path-based vs. distributional information in recognizing lexical semantic relations, *ArXiv Preprint ArXiv:1608.05014*.

T. Siddiqui, X. Ren, A. Parameswaran, and J. Han (2016). FacetGist: Collective extraction of document facets in large technical corpora, in *Proc. of the 25th ACM International Conference on Information and Knowledge Management (CIKM)*. DOI: 10.1145/2983323.2983828. 157

S. Singh, A. Subramanya, F. Pereira, and A. McCallum (2012). Wikilinks: A large-scale cross-document coreference corpus labeled via links to Wikipedia, *UM-CS-2012–015*. 62

R. Snow, D. Jurafsky, and A. Y. Ng (2004). Learning syntactic patterns for automatic hypernym discovery, *Advances in Neural Information Processing Systems 17*. 75, 77

R. Snow, D. Jurafsky, and A. Y. Ng (2005). Learning syntactic patterns for automatic hypernym discovery. in *NIPS*, pp. 1297–1304. 112

N. Srivastava, G. E. Hinton, A. Krizhevsky, I. Sutskever, and R. Salakhutdinov (2014). Dropout: A simple way to prevent neural networks from overfitting, *Journal of Machine Learning Research*, vol. 15, no. 1, pp. 1929–1958. 125

F. M. Suchanek, G. Kasneci, and G. Weikum (2007). Yago: A core of semantic knowledge, in *Proc. of the 16th International Conference on World Wide Web*, ACM, pp. 697–706. DOI: 10.1145/1242572.1242667. 22

A. Sun and R. Grishman (2010). Semi-supervised semantic pattern discovery with guidance from unsupervised pattern clusters, in *Proc. of the 23rd International Conference on Computational Linguistics: Posters*, pp. 1194–1202, Association for Computational Linguistics. 77

H. Sun, H. Ma, W.-t. Yih, C.-T. Tsai, J. Liu, and M.-W. Chang (2015). Open domain question answering via semantic enrichment, in *WWW*. DOI: 10.1145/2736277.2741651. 87

M. Surdeanu, J. Tibshirani, R. Nallapati, and C. D. Manning (2012). Multi-instance multi-label learning for relation extraction, in *EMNLP*. 30, 88

S. Takamatsu, I. Sato, and H. Nakagawa (2012). Reducing wrong labels in distant supervision for relation extraction, in *ACL*. 30

P. P. Talukdar and F. Pereira (2010). Experiments in graph-based semi-supervised learning methods for class-instance acquisition, in *ACL*. 29, 35, 36

J. Tang, M. Qu, M. Wang, M. Zhang, J. Yan, and Q. Mei (2015). Line: Large-scale information network embedding, in *WWW*. DOI: 10.1145/2736277.2741093. 30, 70, 96, 101, 102, 104, 113, 133, 134, 155

J. Tang, M. Qu, and Q. Mei (2015). Pte: Predictive text embedding through large-scale heterogeneous text networks, in *KDD*. DOI: 10.1145/2783258.2783307. 70

F. Tao, H. Zhuang, C. W. Yu, Q. Wang, T. Cassidy, L. Kaplan, C. Voss, and J. Han (2016). Multi-dimensional, phrase-based summarization in text cubes, *Data Engineering*, p. 74. 155

K. Toutanova, D. Klein, C. D. Manning, and Y. Singer (2003). Feature-rich part-of-speech tagging with a cyclic dependency network, in *HLT-NAACL*. DOI: 10.3115/1073445.1073478. 51

K. Toutanova, D. Chen, P. Pantel, P. Choudhury, and M. Gamon (2015). Representing text for joint embedding of text and knowledge bases, in *EMNLP*. DOI: 10.18653/v1/d15-1174. 30, 112, 113, 145

P. Tseng (2001). Convergence of a block coordinate descent method for nondifferentiable minimization, *Journal of Optimization Theory and Applications*, vol. 109, no. 3, pp. 475–494. DOI: 10.1023/a:1017501703105. 50, 68

P. Varma, B. He, D. Iter, P. Xu, R. Yu, C. De Sa, and C. Ré (2016). Socratic learning: Correcting misspecified generative models using discriminative models, *ArXiv Preprint ArXiv:1610.08123*. 119

E. M. Voorhees (1994). Query expansion using lexical-semantic relations, in *Proc. of the 17th Annual International ACM SIGIR Conference on Research and Development in Information Retrieval*, pp. 61–69, Springer-Verlag, Inc., NY. DOI: 10.1007/978-1-4471-2099-5_7. 75

D. Vrandečić and M. Krötzsch (2014). Wikidata: A free collaborative knowledgebase, *Communications of the ACM*, vol. 57, no. 10, pp. 78–85. DOI: 10.1145/2629489. 22

D. Wang, S. Zhu, T. Li, and Y. Gong (2013). Comparative document summarization via discriminative sentence selection, *TKDD*, vol. 6, no. 3, p. 12. DOI: 10.1145/2362383.2362386. 158

Z. Wang, J. Zhang, J. Feng, and Z. Chen (2014). Knowledge graph and text jointly embedding, in *EMNLP*. DOI: 10.3115/v1/d14-1167. 113

H. Wang, F. Tian, B. Gao, J. Bian, and T.-Y. Liu (2015). Solving verbal comprehension questions in IQ test by knowledge-powered word embedding, *ArXiv Preprint ArXiv:1505.07909*. 75, 77

H. Wang, F. Tian, B. Gao, C. Zhu, J. Bian, and T.-Y. Liu (2016). Solving verbal questions in IQ test by knowledge-powered word embedding, in *EMNLP*, pp. 541–550. DOI: 10.18653/v1/d16-1052. 112

J. Weeds, D. Clarke, J. Reffin, D. J. Weir, and B. Keller (2014). Learning to distinguish hypernyms and co-hyponyms, in *COLING*, pp. 2249–2259. 75, 77

R. Weischedel and A. Brunstein (2005). BBN pronoun coreference and entity type corpus, *Linguistic Data Consortium*, vol. 112. 69

R. Weischedel, E. Hovy, M. Marcus, M. Palmer, R. Belvin, S. Pradhan, L. Ramshaw (2011). and N. Xue, Ontonotes: A large training corpus for enhanced processing. 69

R. West, E. Gabrilovich, K. Murphy, S. Sun, R. Gupta, and D. Lin (2014). Knowledge base completion via search-based question answering, in *WWW*. DOI: 10.1145/2566486.2568032. 87

J. Weston, S. Bengio, and N. Usunier (2011). Wsabie: Scaling up to large vocabulary image annotation, in *IJCAI*. DOI: 10.5591/978-1-57735-516-8/IJCAI11-460. 65, 67

F. Wu and D. S. Weld (2010). Open information extraction using Wikipedia, in *Proc. of the 48th Annual Meeting of the Association for Computational Linguistics*, pp. 118–127. 75

P. Xie, D. Yang, and E. P. Xing (2015). Incorporating word correlation knowledge into topic modeling, in *HLT-NAACL*. DOI: 10.3115/v1/n15-1074. 75

W. Xu, R. Hoffmann, L. Zhao, and R. Grishman (2013). Filling knowledge base gaps for distant supervision of relation extraction, in *ACL*. 110

C. Xu, Y. Bai, J. Bian, B. Gao, G. Wang, X. Liu, and T.-Y. Liu (2014). Rc-net: A general framework for incorporating knowledge into word representations, in *Proc. of the 23rd ACM International Conference on Conference on Information and Knowledge Management*, pp. 1219–1228, ACM. DOI: 10.1145/2661829.2662038. 112, 113

Y. Xu, L. Mou, G. Li, Y. Chen, H. Peng, and Z. Jin (2015). Classifying relations via long short term memory networks along shortest dependency paths. in *EMNLP*, pp. 1785–1794. DOI: 10.18653/v1/d15-1206. 112

M. Yahya, S. Whang, R. Gupta, and A. Y. Halevy (2014). Renoun: Fact extraction for nominal attributes, in *EMNLP*. DOI: 10.3115/v1/d14-1038. 112, 139

D. Yogatama, D. Gillick, and N. Lazic (2015). Embedding methods for fine grained entity type classification, in *ACL*. DOI: 10.3115/v1/p15-2048. 30, 60, 64, 69, 70, 104

M. A. Yosef, S. Bauer, J. Hoffart, M. Spaniol, and G. Weikum (2012). Hyena: Hierarchical type classification for entity names, in *COLING*. 59, 62, 68, 70, 91, 104

X. Yu, X. Ren, Y. Sun, Q. Gu, B. Sturt, U. Khandelwal, B. Norick, and J. Han (2014). Personalized entity recommendation: A heterogeneous information network approach, in *Proc. of the 7th ACM International Conference on Web Search and Data Mining (WSDM)*. DOI: 10.1145/2556195.2556259. 59

D. Yu and H. Ji (2016). Unsupervised person slot filling based on graph mining, in *ACL*. DOI: 10.18653/v1/p16-1005. 139

Q. T. Zeng, D. Redd, T. C. Rindflesch, and J. R. Nebeker (2012). Synonym, topic model and predicate-based query expansion for retrieving clinical documents, in *AMIA*. 75

D. Zeng, K. Liu, Y. Chen, and J. Zhao (2015). Distant supervision for relation extraction via piecewise convolutional neural networks. in *EMNLP*, pp. 1753–1762. DOI: 10.18653/v1/d15-1203. 112

W. Zeng, Y. Lin, Z. Liu, and M. Sun (2017). Incorporating relation paths in neural relation extraction, in *EMNLP*. DOI: 10.18653/v1/d17-1186.

C. Zhai, A. Velivelli, and B. Yu (2004). A cross-collection mixture model for comparative text mining, in *SIGKDD*. DOI: 10.1145/1014052.1014150. 158

L. Zhang, L. Li, C. Shen, and T. Li (2015). Patentcom: A comparative view of patent document retrieval, *SDM*. DOI: 10.1137/1.9781611974010.19. 158

Authors' Biographies

XIANG REN

Xiang Ren is an Assistant Professor in the Department of Computer Science at USC, affiliated faculty at USC ISI, and a part-time data science advisor at Snap Inc. At USC, Xiang is part of the Machine Learning Center, NLP community, and Center on Knowledge Graphs. Prior to that, he was a visiting researcher at Stanford University, and received his Ph.D. in CS@UIUC. His research develops computational methods and systems that extract machine-actionable knowledge from massive unstructured data (e.g., text data), and particular focuses on problems in the space of modeling sequence and graph data under weak supervision (learning with partial/noisy labels, and semi-supervised learning) and indirect supervision (multi-task learning, transfer learning, and reinforcement learning). Xiang's research has been recognized with several prestigious awards including a Yahoo!-DAIS Research Excellence Award, a Yelp Dataset Challenge award, a C. W. Gear Outstanding Graduate Student Award and a David J. Kuck Outstanding M.S. Thesis Award. Technologies he developed have been transferred to U.S. Army Research Lab, National Institute of Health, Microsoft, Yelp, and TripAdvisor.

JIAWEI HAN

Jiawei Han is the Abel Bliss Professor in the Department of Computer Science, University of Illinois at Urbana-Champaign. He has been researching into data mining, information network analysis, database systems, and data warehousing, with over 900 journal and conference publications. He has chaired or served on many program committees of international conferences in most data mining and database conferences. He also served as the founding Editor-In-Chief of *ACM Transactions on Knowledge Discovery from Data* and the Director of Information Network Academic Research Center supported by U.S. Army Research Lab (2009–2016), and is the co-Director of KnowEnG, an NIH funded Center of Excellence in Big Data Computing since 2014. He is a Fellow of ACM, a Fellow of IEEE, and received 2004 ACM SIGKDD Innovations Award, 2005 IEEE Computer Society Technical Achievement Award, and 2009 M. Wallace McDowell Award from IEEE Computer Society. His co-authored book *Data Mining: Concepts and Techniques* has been adopted as a popular textbook worldwide.